Instructor's Solutio

to accompany

Introduction to Flight

Fourth Edition

John D. Anderson, Jr.
Curator for Aerodynamics
National Air and Space Museum
and
Professor Emeritus
University of Maryland

Boston   Burr Ridge, IL   Dubuque, IA   Madison, WI   New York   San Francisco   St. Louis
Bangkok   Bogotá   Caracas   Lisbon   London   Madrid
Mexico City   Milan   New Delhi   Seoul   Singapore   Sydney   Taipei   Toronto

McGraw-Hill Higher Education
A Division of The McGraw·Hill Companies

Instructor's Solutions Manual to accompany
INTRODUCTION TO FLIGHT

Copyright © 2000, 1989, 1985, 1978 by The McGraw-Hill Companies, Inc. All rights reserved.
Printed in the United States of America.
The contents of, or parts thereof, may be reproduced for use with
INTRODUCTION TO FLIGHT
John D. Anderson, Jr.
provided such reproductions bear copyright notice and may not be reproduced in
any form for any other purpose without permission of the publisher.

1 2 3 4 5 6 7 8 9 0 HAM/HAM 9 0 9 8 7 6 5 4 3 2 1 0 9

ISBN 0-07-109283-8

http://www.mhhe.com

CONTENTS

Chapter		Page
1	The First Aeronautical Engineers	
2	Fundamental Thoughts	1
3	The Standard Atmosphere	6
4	Basic Aerodynamics	11
5	Airfoils, Wings, and Other Aerodynamic Shapes	38
6	Elements of Airplane Performance	61
7	Principles of Stability and Control	86
8	Astronautics	93
9	Propulsion	99
10	Flight Vehicle Structures and Materials	113
11	Hypersonic Vehicles	116

2.1 $\rho = p/RT = (1.2)(1.01 \times 10^5)/(287)(300)$

$\rho = 1.41 \text{ kg/m}^2$

$v = 1/\rho = 1/1.41 = \boxed{0.71 \text{ m}^3/\text{kg}}$

2.2 Mean kinetic energy of each atom =

$\frac{3}{2} kT = \frac{3}{2}(1.38 \times 10^{-23})(500) = 1.035 \times 10^{-20}$ J

One kg-mole, which has a mass of 4 kg, has 6.02×10^{26} atoms. Hence 1 kg has $\frac{1}{4}$ (6.02×10^{26}) = 1.505×10^{26} atoms.

Total internal energy = (energy per atom)(number of atoms)

$= (1.035 \times 10^{-20})(1.505 \times 10^{26}) = \boxed{1.558 \times 10^6 \text{ J}}$

2.3 $\rho = \dfrac{p}{RT} = \dfrac{2116}{(1716)(460+59)} = 0.00237 \dfrac{\text{slug}}{\text{ft}^3}$

Volume of the room = (20)(15)(8) = 2400 ft³

Total mass in the room = (2400)(0.00237) = 5.688 slug

Weight = (5.688)(32.2) = $\boxed{183 \text{ lb}}$

2.4 $\rho = \dfrac{p}{RT} = \dfrac{2116}{(1716)(460-10)} = 0.00274 \dfrac{\text{slug}}{\text{ft}^3}$

Since the volume of the room is the same, we can simply compare densities between the two problems.

$$\Delta\rho = 0.00274 - 0.00237 = 0.00037 \frac{\text{slug}}{\text{ft}^3}$$

$$\% \text{ change} = \frac{\Delta\rho}{\rho} = \frac{0.00037}{0.00237} \times (100) = \boxed{15.6\% \text{ increase}}$$

2.5 First, calculate the density from the known mass and volume,

$$\rho = 1500/900 = 1.67 \text{ lb}_m/\text{ft}^3$$

In consistent units, $\rho = 1.67/32.2 = 0.052 \text{ slug/ft}^3$. Also, $T = 70F = 70 + 460 = 530R$.

Hence,

$$p = \rho RT = (0.52)(1716)(530)$$

$$p = 47290 \text{ lb/ft}^2$$

or $p = 47290/2116 = \boxed{22.3 \text{ atm}}$

2.6 $p = \rho RT$

$\ell n p = \ell n \rho + \ell n R + \ell n T$

Differentiating with respect to time,

$$\frac{1}{p}\frac{dp}{dt} = \frac{1}{\rho}\frac{d\rho}{dt} + \frac{1}{T}\frac{dT}{dt}$$

or, $$\frac{dp}{dt} = \frac{p}{\rho}\frac{d\rho}{dt} + \frac{p}{T}\frac{dT}{dt}$$

or, $$\frac{dp}{dt} = RT\frac{d\rho}{dt} + \rho R\frac{dT}{dt} \qquad (1)$$

At the instant there is 1000 lb_m of air in the tank, the density is

$$\rho = 1000/900 = 1.11 \text{ lb}_m/\text{ft}^3$$

$\rho = 1.11/32.2 = 0.0345$ slug/ft^3

Also, in consistent units, is is given that

$T = 50 + 460 = 510$R

and that

$$\frac{dT}{dt} = 1\text{F/min} = 1\text{R/min} = 0.0167\text{R/sec}$$

From the given pumping rate, and the fact that the volume of the tank is 900 ft^3, we also have

$$\frac{d\rho}{dt} = \frac{0.5 \text{ lb}_m/\text{sec}}{900 \text{ ft}^3} = 0.000556 \text{ lb}_m/(\text{ft}^3)(\text{sec})$$

$$\frac{d\rho}{dt} = \frac{0.000556}{32.2} = 1.73 \times 10^{-5} \text{ slug}/(\text{ft}^3)(\text{sec})$$

Thus, from equation (1) above,

$$\frac{dp}{dt} = (1716)(510)(1.73 \times 10^{-5}) + (0.0345)(1716)(0.0167)$$

$$= 15.1 + 0.99 = 16.1 \text{ lb}/(\text{ft}^2)(\text{sec}) = \frac{16.1}{2116}$$

$$= \boxed{0.0076 \text{ atm/sec}}$$

2.7 In consistent units,

$T = -10 + 273 = 263$K

Thus,

$\rho = p/RT = (1.7 \times 10^4)/(287)(263)$

$\rho = \boxed{0.225 \text{ kg/m}^3}$

2.8 $\rho = p/RT = 0.5 \times 10^5/(287)(240) = 0.726$ kg/m³

$v = 1/\rho = 1/0.726 = \boxed{1.38 \text{ m}^3/\text{kg}}$

2.9

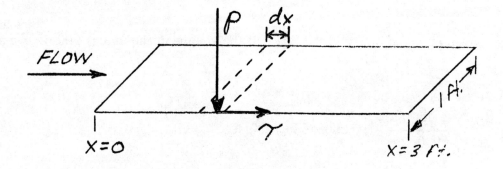

F_p = Force due to pressure = $\int_0^3 p \, dx = \int_0^3 (2116 - 10x) \, dx$

$= [2116x - 5x^2]_0^3 = 6303$ lb perpendicular to wall.

F_τ = Force due to shear stress = $\int_0^3 \tau \, dx = \int_0^3 \dfrac{90}{(x+9)^{\frac{1}{2}}} dx$

$= [180(x+9)^{\frac{1}{2}}]_0^3 = 623.5 - 540 = 83.5$ lb tangential to wall.

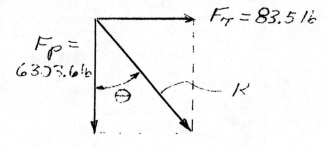

Magnitude of the resultant aerodynamic force =

$R = \sqrt{(6303)^2 + (83.5)^2} = \boxed{6303.6 \text{ lb}}$

$$\theta = \text{Arc Tan}\left(\frac{83.5}{6303}\right) = \boxed{0.76°}$$

2.10 $V = \dfrac{3}{2} V_\infty \sin\theta$

Minimum velocity occurs when $\sin\theta = 0$, i.e. when $\theta = 0°$ and $180°$.

$\boxed{V_{min} = 0}$ at $\theta = 0°$ and $180°$, i.e., at its most forward and rearward points.

Maximum velocity occurs when $\sin\theta = 1$, i.e. when $\theta = 90°$. Hence

$$V_{max} = \frac{3}{2}(85)(1) = \boxed{127.5 \text{ mph}} \text{ at } \theta = 90°,$$

i.e., the entire rim of the sphere in a plane perpendicular to the freestream direction.

2.11 The mass of air displaced is

$$M = (2.2)(0.002377) = 5.23 \times 10^{-3} \text{ slug}$$

The weight of this air is

$$W_{air} = (5.23 \times 10^{-3})(32.2) = 0.168 \text{ lb}$$

This is the lifting force on the balloon due to the outside air. However, the helium inside the balloon has weight, acting in the downward direction. The weight of the helium is less than that of air by the ratio of the molecular weights

$$W_{H_e} = (0.168)\frac{4}{28.8} = 0.0233 \text{ lb}.$$

Hence, the maximum weight that can be lifted by the balloon is

$$0.168 - 0.0233 = \boxed{0.145 \text{ lb}}$$

3.1 An examination of the standard temperature distribution through the atmosphere given in Figure 3.3 of the text shows that both 12 km and 18 km are in the same constant temperature region. Hence, the equations that apply are Eqs. (3.9) and (3.10) in the text. Since we are in the same isothermal region with therefore the same base values of p and ρ, these equations can be written as

$$\frac{\rho_2}{\rho_1} = \frac{p_2}{p_1} = e^{-(g_o/RT)(h_2-h_1)}$$

where points 1 and 2 are any two arbitrary points in the region. Hence, with $g_o = 9.8$ m/sec² and $R = 287$ joule/kgK, and letting points 1 and 2 correspond to 12 km and 18 km altitudes respectively,

$$\frac{\rho_2}{\rho_1} = \frac{p_2}{p_1} = e^{-\frac{9.8}{(287)(216.66)}(6000)} = 0.3884$$

Hence:

$$p_2 = (0.3884)(1.9399 \times 10^4) = \boxed{7.53 \times 10^3 \text{ N/m}^2}$$

$$\rho_2 = (0.3884)(3.1194 \times 10^{-1}) = \boxed{0.121 \text{ kg/m}^3}$$

and of course

$$T_2 = \boxed{216.66 \text{K}}$$

These answers check the results listed in Appendix A of the text within round-off error.

3.2 From Appendix A of the text, we see immediately that $p = 2.65 \times 10^4$ N/m² corresponds to 10,000 m, or 10 km, in the standard atmosphere. Hence,

$$\boxed{\text{pressure altitude} = 10 \text{ km}}$$

The outside air density is

$$\rho = \frac{p}{RT} = \frac{2.65 \times 10^4}{(287)(220)} = 0.419 \text{ kg/m}^3$$

From Appendix A, this value of ρ corresponds to 9.88 km in the standard atmosphere. Hence,

$$\boxed{\text{density altitude} = 9.88 \text{ km}}$$

3.3 At 35,000 ft, from Appendix B, we find that $p = 4.99 \times 10^2 = \boxed{499 \text{ lb/ft}^2}$.

3.4 From Appendix B in the text,

33,500 ft corresponds to $p = 535.89$ lb/ft^2

32,000 ft corresponds to $\rho = 8.2704 \times 10^{-4}$ slug/ft^3

Hence,

$$T = \frac{p}{\rho R} = \frac{535.89}{(8.2704 \times 10^{-4})(1716)} = \boxed{378 \text{ R}}$$

3.5 $\dfrac{|h - h_G|}{h} = 0.02 = \left|1 - \dfrac{h_G}{h}\right|$

From Eq. (3.6), the above equation becomes

$$\left|1 - \left(\frac{r + h_G}{r}\right)\right| = \left|1 - 1 - \frac{h_G}{r}\right| = 0.02$$

$$h_G = 0.02\, r = 0.02\, (6.357 \times 10^6)$$

$$h_G = 1.27 \times 10^5 \text{ m} = \boxed{127 \text{ km}}$$

3.6 $T = 15 - 0.0065h = 15 - 0.0065(5000) = -17.5°C = 255.5°K$

$$a = \frac{dT}{dh} = -0.0065$$

From Eq. (3.12)

$$\frac{p}{p_1} = \left(\frac{T}{T_1}\right)^{-g_o/aR} = \left(\frac{255.5}{288}\right)^{-(9.8)/(-0.0065)(287)} = 0.533$$

$p = 0.533\, p_1 = 0.533\,(1.01 \times 10^5) = \boxed{5.38 \times 10^4 \text{ N/m}^2}$

3.7 $\ln \dfrac{p}{p_1} = -\dfrac{g}{RT}(h - h_1)$

$$h - h_1 = -\frac{1}{g} RT \ln \frac{p}{p_1} = -\frac{1}{24.9}(4157)(150)\ln 0.5$$

Letting $h_1 = 0$ (the surface)

$h = 17{,}358 \text{ m} = \boxed{17.358 \text{ km}}$

3.8   A standard altitude of 25,000 ft falls within the first gradient region in the standard atmosphere. Hence, the variation of pressure and temperature are given by:

$$\frac{p}{p_1} = \left(\frac{T}{T_1}\right)^{-\frac{g}{aR}} \tag{1}$$

and

$$T = T_1 + a(h - h_1) \tag{2}$$

Differentiating Eq. (1) with respect to time:

$$\frac{1}{p_1}\frac{dp}{dt} = \left(\frac{1}{T_1}\right)^{-\frac{g}{aR}}\left(-\frac{g}{AR}\right) T^{\left(-\frac{g}{aR}-1\right)} \frac{dT}{dt} \qquad (3)$$

Differentiating Eq. (2) with respect to time:

$$\frac{dT}{dt} = a\frac{dh}{dt} \qquad (4)$$

Substitute Eq. (4) into (3)

$$\frac{dp}{dt} = -p_1(T_1)^{\frac{g}{aR}}\left(\frac{g}{R}\right) T^{-\left(\frac{g}{aR}+1\right)} \frac{dh}{dt} \qquad (5)$$

In Eq. (5), dh/dt is the rate-of-climb, given by dh/dt = 500 ft/sec. Also, in the first gradient region, the lapse rate can be calculated from the tabulations in Appendix B. For example, take 0 ft and 10,000 ft, we find

$$a = \frac{T_2 - T_1}{h_2 - h_1} = \frac{483.04 - 518.69}{10,000 - 0} = -0.00357 \frac{°R}{ft}$$

Also from Appendix B, p_1 = 2116.2 lb/ft² at sea level, and T = 429.64 °R at 25,000 ft. Thus,

$$\frac{g}{aR} = \frac{32.2}{(-0.00357)(1716)} = -5.256$$

Hence, from Eq. (5)

$$\frac{dp}{dt} = -(2116.2)(518.69)^{-5.256}\left(\frac{32.2}{1716}\right)(429.64)^{4.256}(500)$$

$$\boxed{\frac{dp}{dt} = -17.17 \frac{lb}{ft^2 \, sec}}$$

3.9 From the hydrostatic equation, Eq. (3.2) or (3.3),

$$dp = -\rho g_o \, dh$$

or $$\frac{dp}{dt} = -\rho g_o \frac{dh}{dt}$$

The upward speed of the elevator is dh/dt, which is

$$\frac{dh}{dt} = \frac{dp/dt}{-\rho g_o}$$

At sea level, $\rho = 1.225$ kg/m^3. Also, a one-percent change in presure per minute starting from sea level is

$$\frac{dp}{dt} = -(1.01 \times 10^5)(0.01) = -1.01 \times 10^3 \text{ N/m}^2 \text{ per minute}$$

Hence

$$\frac{dh}{dt} = \frac{-1.01 \times 10^3}{(1.225)(9.8)} = \boxed{84.1 \text{ meter per minute}}$$

3.10 From Appendix B:

At 35,500 ft: p = 535.89 lb/ft^2

At 34,000 ft: p = 523.47 lb/ft^2

For a pressure of 530 lb/ft^2, the pressure altitude is

$$33,500 + 500 \left(\frac{535.89 - 530}{535.89 - 523.47}\right) = \boxed{33737 \text{ ft}}$$

The density at the altitude at which the airplane is flying is

$$\rho = \frac{p}{RT} = \frac{530}{(1716)(390)} = 7.919 \times 10^{-4} \text{ slug/ft}^3$$

From Appendix B:

At 33,000 ft: $\rho = 7.9656 \times 10^{-4}$ slug/ft^3

At 33,500 ft: $\rho = 7.8165 \times 10^{-4}$ slug/ft^3

Hence, the density altitude is

$$33,000 + 500 \left(\frac{7.9656 - 7.919}{7.9656 - 7.8165}\right) = \boxed{33,156 \text{ ft}}$$

4.1 $A_1 V_1 = A_2 V_2$

Let points 1 and 2 denote the inlet and exit conditions respectively. Then,

$$V_2 = V_1 \left(\frac{A_1}{A_2}\right) = (5)\left(\frac{1}{4}\right) = \boxed{1.25 \text{ ft/sec}}$$

4.2 From Bernoulli's equation,

$$p_1 + \rho \frac{V_1^2}{2} = p_2 + \rho \frac{V_2^2}{2}$$

$$p_2 - p_1 = \frac{\rho}{2}(V_1^2 - V_2^2)$$

In consistent units,

$$\rho = \frac{62.4}{32.2} = 1.94 \text{ slug/ft}^3$$

Hence,

$$p_2 - p_1 = \frac{1.94}{2}[(5)^2 - (1.25)^2]$$

$$p_2 - p_1 = 0.97(23.4) = \boxed{22.7 \text{ lb/ft}^2}$$

4.3 From Appendix A; at 3000m altitude,

$p_1 = 7.01 \times 10^4$ N/m^2

$\rho = 0.909$ kg/m^3

From Bernoulli's equation,

$$p_2 = p_1 + \frac{\rho}{2}(V_1^2 - V_2^2)$$

$$p_2 = 7.01 \times 10^4 + \frac{0.909}{2}[60^2 - 70^2]$$

$$p_2 = 7.01 \times 10^4 - 0.059 \times 10^4 = \boxed{6.95 \times 10^4 \text{ N/m}^2}$$

4.4 From Bernoulli's equation,

$$p_1 + \frac{\rho}{2}V_1^2 = p_2 + \frac{\rho}{2}V_2^2$$

Also from the incompressible continuity equation

$$V_2 = V_1 (A_1/A_2)$$

Combining,

$$p_1 + \frac{\rho}{2}V_1^2 = p_2 + \frac{\rho}{2}(A_1/A_2)^2$$

$$V_1 = \sqrt{\frac{2(p_1 - p_2)}{\rho\left[(A_1/A_2)^2 - 1\right]}}$$

At standard sea level, $\rho = 0.002377$ slug/ft^3. Hence,

$$V_1 = \sqrt{\frac{2(80)}{(.002377)[(4)^2 - 1]}} = \boxed{67 \text{ ft/sec}}$$

Note that also $V_1 = 67\left(\dfrac{60}{80}\right) = 46$ mi/h. (This is approximately the landing speed of World War I vintage aircraft).

4.5 $p_1 + \dfrac{1}{2}\rho V_1^2 = p_3 + \dfrac{1}{2}\rho V_3^2$

$$V_1^2 = \dfrac{2(p_3 - p_1)}{\rho} + V_3^2 \tag{1}$$

$$A_1 V_1 = A_3 V_3, \text{ or } V_3 = \dfrac{A_1}{A_3} V_1 \tag{2}$$

Substitute (2) into (1)

$$V_1^2 = \dfrac{2(p_3 - p_1)}{\rho} + \left(\dfrac{A_1}{A_3}\right)^2 V_1^2$$

or, $\quad V_1 = \sqrt{\dfrac{2(p_3 - p_1)}{\rho\left[1 - \left(\dfrac{A_1}{A_3}\right)^2\right]}} \tag{3}$

Also,

$$A_1 V_1 = A_2 V_2$$

or, $\quad V_2 = \left(\dfrac{A_1}{A_2}\right) V_1 \tag{4}$

Substitute (3) into (4)

$$V_1 = \dfrac{A_1}{A_2}\sqrt{\dfrac{2(p_3 - p_1)}{\rho\left[1 - \left(\dfrac{A_1}{A_3}\right)^2\right]}}$$

$$V_2 = \frac{3}{1.5} \sqrt{\frac{2(1.00-1.02) \times 10^5}{(1.225)\left[1-\left(\frac{3}{2}\right)^2\right]}}$$

$$V_2 = \boxed{102.22 \text{ m/sec}}$$

Note: It takes a pressure difference of <u>only</u> 0.02 atm to produce such a high velocity.

4.6 $V_1 = 130 \text{ mph} = 130 \left(\frac{88}{60}\right) = 190.7 \text{ ft/sec}$

$$p_1 + \frac{1}{2}\rho V_1^2 = p_2 + \frac{1}{2}\rho V_2^2$$

$$V_2^2 = \frac{2}{\rho}(p_1 - p_2) + V_1^2$$

$$V_2^2 = \frac{2(1760.9 - 1750.0)}{0.0020482} + (190.7)^2$$

$$V_2 = \boxed{216.8 \text{ ft/sec}}$$

4.7   From Bernoulli's equation,

$$p_1 - p_2 = \frac{\rho}{2}(V_2^2 - V_1^2)$$

And from the incompressible continuity equation,

$$V_2 = V_1 (A_1/A_2)$$

Combining:

$$p_1 - p_2 = \frac{\rho}{2} V_1^2 \left[(A_1/A_2)^2 - 1\right]$$

Hence, the maximum pressure difference will occur when simultaneously:

1. V_1 is maximum

2. ρ is maximum i.e. sea level

The design maximum velocity is 90 m/sec, and $\rho = 1.225$ kg/m³ at sea level. Hence,

$$p_1 - p_2 = \frac{1.225}{2}(90)^2[(1.3)^2 - 1] = \boxed{3423 \text{ N/m}^2}$$

Please note: In reality the airplane will most likely exceed 90 m/sec in a dive, so the airspeed indicator should be designed for a maximum velocity somewhat above 90 m/sec.

4.8 The isentropic relations are

$$\frac{p_e}{p_o} = \left(\frac{\rho_e}{\rho_o}\right)^\gamma = \left(\frac{T_e}{T_o}\right)^{\frac{\gamma}{\gamma-1}}$$

Hence,

$$T_e = T_o \left(\frac{p_e}{p_o}\right)^{\frac{\gamma}{\gamma-1}} = (300)\left(\frac{1}{10}\right)^{\frac{1.4-1}{1.4}} = \boxed{155 K}$$

From the equation of state:

$$\rho_o = \frac{p_o}{RT_o} = \frac{(10)(1.01 \times 10^5)}{(287)(300)} = 11.73 \text{ kg/m}^3$$

Thus,

$$\rho_e = \rho_o \left(\frac{p_e}{p_o}\right)^{\frac{1}{\gamma}} = 11.73 \left(\frac{1}{10}\right)^{\frac{1}{1.4}} = \boxed{2.26 \text{ kg/m}^3}$$

As a check on the results, apply the equation of state at the exit.

$p_e = \rho_e RT_e$?

$1.01 \times 10^5 = (2.26)(287)(155)$

$1.01 \times 10^5 = 1.01 \times 10^5$ It checks!

4.9 Since the velocity is essentially zero in the reservoir, the energy equation written between the reservoir and the exit is

$$h_o = h_e + \frac{V_e^2}{2}$$

or, $V_e^2 = 2(h_o - h_e)$ (1)

However, $h = c_p T$. Thus Eq. (1) becomes

$$V_e^2 = 2 c_p (T_o - T_e)$$

$$V_e^2 = 2 c_p T_o \left(1 - \frac{T_e}{T_o}\right) \quad (2)$$

However, the flow is isentropic, hence

$$\frac{T_e}{T_o} = \left(\frac{p_e}{p_o}\right)^{\frac{\gamma-1}{\gamma}} \quad (3)$$

Substitute (3) into (1).

$$V_e = \sqrt{2 c_p T_o \left[1 - \left(\frac{p_e}{p_o}\right)^{\frac{\gamma-1}{\gamma}}\right]} \quad (4)$$

This is the desired result. Note from Eq. (4) that V_e increases as T_o increases, and as p_e/p_o decreases. Equation (4) is a useful formula for rocket engine performance analysis.

4.10 The flow velocity is certainly large enough that the flow must be treated as compressible. From the energy equation,

$$c_p T_1 + \frac{V_1^2}{2} = c_p T_2 + \frac{V_2^2}{2} \tag{1}$$

At a standard altitude of 5 km, from Appendix A,

$p_1 = 5.4 \times 10^4 \text{ N/m}^2$

$T_1 = 255.7 \text{ K}$

Also, for air, $c_p = 1005$ joule/(kg)(K). Hence, from Eq. (1) above,

$$T_2 = T_1 + \frac{V_1^2 - V_2^2}{2 c_p}$$

$$T_2 = 255.7 + \frac{(270)^2 - (330)^2}{2(1005)}$$

$T_2 = 255.7 - 17.9 = 237.8 \text{K}$

Since the flow is also isentropic,

$$\frac{p_2}{p_1} = \left(\frac{T_2}{T_1}\right)^{\frac{\gamma-1}{\gamma}}$$

Thus,

$$p_2 = p_1 \left(\frac{T_2}{T_1}\right)^{\frac{\gamma-1}{\gamma}} = 5.4 \times 10^4 \left(\frac{237.8}{255.7}\right)^{\frac{1.4}{1.4-1}}$$

$p_2 = \boxed{4.19 \times 10^4 \text{ N/m}^2}$

Please note: This problem and problem 4.3 ask the same question. However, the flow velocities in the present problem require a compressible analysis. Make certain to examine the solutions of both problems 4.10 and 4.3 in order to contrast compressible versus incompressible analyses.

4.11 From the energy equation

$$c_p T_o = c_p T_e + \frac{V_e^2}{2}$$

or, $T_e = T_o - \dfrac{V_e^2}{2 c_p}$

$$T_e = 1000 - \frac{1500^2}{2(6000)} = 812.5 R$$

In the reservoir, the density is

$$\rho_o = \frac{p_o}{RT_o} = \frac{(7)(2116)}{(1716)(1000)} = 0.0086 \text{ slug/ft}^3$$

From the isentropic relation,

$$\frac{\rho_e}{\rho_o} = \left(\frac{T_e}{T_o}\right)^{\frac{1}{\gamma-1}}$$

$$\rho_e = 0.0086 \left(\frac{812.5}{1000}\right)^{\frac{1}{1.4-1}} = 0.0051 \text{ slug/ft}^3$$

From the continuity equation,

$$\dot{m} = \rho_e A_e V_e$$

Thus, $A_e = \dfrac{\dot{m}}{\rho_e V_e}$

In consistent units,

$$\dot{m} = \frac{1.5}{32.2} = 0.047 \text{ slug/sec.}$$

Hence,

$$A_e = \frac{\dot{m}}{\rho_e V_e} = \frac{0.047}{(0.0051)(1500)} = \boxed{0.0061 \text{ ft}^2}$$

4.12 $V_1 = 1500 \text{ mph} = 1500 \left(\frac{88}{60}\right) = 2200 \text{ ft/sec}$

$$C_p T_1 + \frac{V_1^2}{2} C_p T_2 \frac{V_2^2}{2}$$

$$V_2^2 = 2 C_p (T_1 - T_2) + V_1^2$$

$$V_2^2 = 2 (6000)(389.99 - 793.32) + (2200)^2$$

$$V_2 = \boxed{6.3 \text{ ft/sec}}$$

Note: This is a very <u>small</u> velocity compared to the initial freestream velocity of 2200 ft/sec. At the point in question, the velocity is very near zero, and hence the point is nearly a stagnation point.

4.13 At the inlet, the mass flow of air is

$$\dot{m}_{air} = \rho A V = (3.6391 \times 10^{-4})(20)(2200) = 16.0/\text{slug/sec}$$

$$\dot{m}_{fuel} = (0.05)(16.01) = 0.8 \text{ slug/sec}$$

Total mass flow at exit = $16.01 + 0.8 = \boxed{16.81 \text{ slug/sec}}$

4.14 From problem 4.11,

$V_e = 1500 \text{ ft/sec}$

$T_e = 812.5 R$

Hence,

$$a_e = \sqrt{\gamma RT_e} = \sqrt{(1.4)(1716)(812.5)}$$

$$= 1397 \text{ ft/sec}$$

Thus, $M_e = \dfrac{V_e}{a_e} = \dfrac{1500}{1397} = \boxed{1.07}$

Note that the nozzle of problem 4.11 is just barely supersonic.

4.15 From Appendix A,

$T_\infty = 216.66 \text{K}$

Hence,

$$a_\infty = \sqrt{\gamma RT}$$

$$= \sqrt{(1.4)(287)(216.66)} = 295 \text{ m/sec}$$

Thus, $M_\infty = \dfrac{V_\infty}{a_\infty} = \dfrac{250}{295} = \boxed{0.847}$

4.16 At standard sea level, $T_\infty = 518.69 \text{R}$

$$a_\infty = \sqrt{\gamma RT} = \sqrt{(1.4)(1716)(518.69)} = 1116 \text{ ft/sec}$$

$$V_\infty = M_\infty a_\infty = (3)(1116) = 3348 \text{ ft/sec}$$

Since 60 mi/hr - 88 ft/sec., then

$$V_\infty = 3348 \,(60/88) = \boxed{2283 \text{ mi/h}}$$

4.17 V = 2200 ft/sec

$$a = \sqrt{\gamma RT} = \sqrt{(1.4)(1716)(389.99)} = 967.94 \text{ ft/sec}$$

$$M = \frac{V}{a} = \frac{2200}{967.94} = \boxed{2.27}$$

4.18 The test section density is

$$\rho = \frac{p}{RT} = \frac{1.01 \times 10^5}{(287)(300)} = 1.173 \text{ kg/m}^3$$

Since the flow is low speed, consider it to be incompressible, i.e., with the above density throughout.

$$p_1 - p_2 = \frac{\rho}{2} V_2^2 [1 - (A_2/A_1)^2] \qquad (1)$$

In terms of the manometer reading,

$$p_1 - p_2 = \omega \Delta h \qquad (2)$$

where $\omega = 1.33 \times 10^5$ N/m³ for mercury.

Thus, combining Eqs. (1) and (2),

$$\Delta h = \frac{\rho}{2\omega} V_2^2 [1 - (A_2/A_1)^2]$$

$$= \frac{1.173}{(2)(1.33 \times 10^5)} (80)^2 [1 - (1/20)^2]$$

$$\Delta h = 0.028 \text{m} = \boxed{2.8 \text{ cm}}$$

4.19 $V_2 = 200$ mph $= 300 \left(\frac{88}{60}\right) = 293.3$ ft/sec

(a) $p_1 + \frac{1}{2}\rho V_1^2 = p_2 + \frac{1}{2}\rho V_2^2$

$A_1 V_1 = A_2 V_2$

$V_1 = \frac{A_2}{A_1} V_2$

$p_1 + \frac{1}{2}\rho \left(\frac{A_2}{A_1}\right)^2 V_2^2 = p_2 + \frac{1}{2}\rho V_2^2$

$p_1 - p_2 = \frac{1}{2}\rho \left[1 - \left(\frac{A_2}{A_1}\right)^2\right] V_2^2$

$p_1 - p_2 = \frac{0.002377}{2}\left[1 - \left(\frac{4}{20}\right)^2\right](293.3)^2$

$p_1 - p_2 = \boxed{98.15 \text{ lb/ft}^2}$

(b) $p_1 + \frac{1}{2}\rho V_1^2 = p_3 + \frac{1}{2}\rho V_3^3$

$A_1 V_1 = A_2 V_2 \quad : \quad V_1 = \frac{A_2}{A_1} V_2$

$A_2 V_2 = A_3 V_3 \quad : \quad V_3 = \frac{A_2}{A_3} V_2$

$p_1 - p_3 = \frac{1}{2}\rho \left[\left(\frac{A_2}{A_3}\right)^2 - \left(\frac{A_2}{A_1}\right)^2\right] V_2^2$

$p_1 - p_3 = \frac{0.002377}{2}\left[\left(\frac{4}{18}\right)^2 - \left(\frac{4}{20}\right)^2\right](293.3)^2$

$p_1 - p_3 = \boxed{0.959 \text{ lb/ft}^2}$

Note: By the addition of a diffuser, the required pressure difference was reduced by an order of magnitude. Since it costs money to produce a pressure difference (say by running compresors or vacuum pumps), then a diffuser, the purpose of which is to improve the aerodynamic efficiency, allows the wind tunnel to be operated more economically.

4.20 In the test section

$$\rho = \frac{p}{RT} = \frac{2116}{(1716)(70+460)} = 0.00233 \text{ slug/ft}^3$$

The flow velocity is low enough so that incompressible flow can be assumed. Hence, from Bernoulli's equation,

$$p_o = p + \frac{1}{2} \rho V^2$$

$$p_o = 2116 + \frac{1}{2}(0.00233)[150(88/60)]^2$$

(Remember that 88 ft/sec = 60 mi/h.)

$$p_o = 2116 + \frac{1}{2}(0.00233)(220)^2$$

$$p_o = \boxed{2172 \text{ lb/ft}^2}$$

4.21 The altimeter meaasures pressure altitude. Thus, from Appendix B, p = 1572 lb/ft². The air density is then

$$\rho = \frac{p}{RT} = \frac{1572}{(1716)(500)} = 0.00183 \text{ slug/ft}^3$$

Hence, from Bernoulli's equation,

$$V_{true} = \sqrt{\frac{2(p_o - p)}{\rho}} = \sqrt{\frac{2(1650 - 1572)}{0.00183}}$$

$$V_{true} = 292 \text{ ft/sec}$$

The equivalent airspeed is

$$V_e = \sqrt{\frac{2(p_o - p)}{\rho_s}} = \sqrt{\frac{2(1650 - 1572)}{0.002377}}$$

$$V_e = \boxed{256 \text{ ft/sec}}$$

4.22 The altimeter measures pressure altitude. Thus, from Appendix A, $p = 7.95 \times 10^4$ N/m². Hence,

$$\rho = \frac{p}{RT} = \frac{7.95 \times 10^4}{(287)(280)} = 0.989 \text{ kg/m}^3$$

The relation between V_{true} and V_e is

$$V_{true}/V_e = \sqrt{\rho_s/\rho}$$

Hence,

$$V_{true} = 50 \sqrt{(1.225)/0.989} = \boxed{56 \text{ m/sec}}$$

4.23 In the test section,

$$a = \sqrt{\gamma RT} = \sqrt{(1.4)(287)(270)} = 329 \text{ m/sec}$$

$$M = V/a = 250/329 = 0.760$$

$$\frac{p_o}{p} = \left(1 + \frac{\gamma - 1}{2} M^2\right)^{\frac{\gamma}{\gamma - 1}} = [1 + 0.2 (0.760)^2]^{3.5} = 1.47$$

Hence,

$$p_o = 1.47p = 1.47 (1.01 \times 10^5) = \boxed{1.48 \times 10^5 \text{ N/m}^2}$$

4.24 $p = 1.94 \times 10^4$ N/m² from Appendix A.

$$M_1^2 = \frac{2}{\gamma - 1}\left[\left(\frac{p_o}{p}\right)^{\frac{\gamma-1}{\gamma}} - 1\right] = \frac{2}{1.4-1}\left[\left(\frac{2.96 \times 10^4}{1.94 \times 10^4}\right)^{0.286} - 1\right]$$

$$M_1^2 = 0.642$$

$$M_1 = \boxed{0.801}$$

4.25 $$\frac{p_o}{p} = \left(1 + \frac{\gamma-1}{2}M^2\right)^{\frac{\gamma}{\gamma-1}}$$

$$\frac{p_o}{p} = [1 + 0.2(0.65)^2]^{3.5} = 1.328$$

$$p = \frac{p_o}{1.328} = \frac{2339}{1.328} = 1761 \text{ lb/ft}^2$$

From Appendix B, this pressure corresponds to a pressure altitude, hence altimeter reading of $\boxed{5000 \text{ ft.}}$

4.26   At standard sea level,

$$T = 518.69 R$$

$$\frac{T_o}{T} = 1 + \frac{\gamma-1}{2}M^2 = 1 + 0.2(0.96)^2 = 1.184$$

$$T_o = 1.184T = 1.184(518.69)$$

$$T_o = 614.3R = 154.3F$$

4.27 $a_1 = \sqrt{\gamma RT_1} = \sqrt{(1.4)(287)(220)} = 297$ m/sec

$M_1 = V_1/a_1 = 596/197 = 2.0$

The flow is supersonic. Hence, the Rayleigh Pitot tube formula must be used.

$$\frac{p_{o_2}}{p_1} = \left[\frac{(\gamma+1)^2 M_1^2}{4\gamma M_1^2 - 2(\gamma-1)}\right]^{\frac{\gamma}{\gamma-1}} \left[\frac{1-\gamma+2\gamma M_1^2}{\gamma+1}\right]$$

$$\frac{p_{o_2}}{p_1} = \left[\frac{(2.4)^2(2)^2}{4(1.4)(2)^2 - 2(0.4)}\right]^{3.5} \left[\frac{1-1.4+2(1.4)(2)^2}{2.4}\right]$$

$$\frac{p_{o_2}}{p_1} = 5.64$$

$p_1 = 2.65 \times 10^4$ N/m^2 from Appendix A.

Hence,

$p_{o_2} = 5.64 (2.65 \times 10^4) = \boxed{1.49 \times 10^5 \text{ N/m}^2}$

4.28 $q = \frac{1}{2} \rho V^2 = \frac{1}{2}\left(\frac{\gamma p}{\gamma p}\right) \rho V^2 = \frac{\gamma}{2} p \left(\frac{\rho}{\gamma p}\right) V^2 = \frac{\gamma}{2} p \frac{V^2}{a^2}$

Hence:

$$\boxed{q = \frac{\gamma}{2} p M^2}$$

4.29 $q_\infty = \frac{\gamma}{2} p_\infty M_\infty^2 = 0.7 p_\infty M_\infty^2$ \hfill (1)

Use Appendix A to obtain the values of p_∞ corresponding to the given values of h. Then use Eq. (1) above to calculate q_∞.

h(km)	60	50	40	30	20
$p_\infty(N/m^2)$	25.6	87.9	299.8	1.19×10^3	5.53×10^3
M	17	9.5	5.5	3	1
$q_\infty(N/M^2)$	5.2×10^3	5.6×10^3	6.3×10^3	7.5×10^3	3.9×10^3

Note that q_∞ progressively increases as the shuttle penetrates deeper into the atmosphere, that it peaks at a slightly supersonic Mach number, and then decreases as the shuttle completes its entry.

4.30 Recall that total pressure is defined as that pressure that would exist if the flow were slowed <u>isentropically</u> to zero velocity. This is a definition; it applies to all flows -- subsonic or supersonic. Hence, Eq. (4.74) applies, no matter whether the flow is subsonic or supersonic.

$$\frac{p_o}{p_\infty} = \left(1 + \frac{\gamma-1}{2} M_\infty^2\right)^{\gamma/(\gamma-1)} = [1 + 0.2\,(2)^2]^{1.4/0.4} = 7.824$$

Hence:

$$p_o = 7.824\, p_\infty = 7.824\,(2116) = \boxed{1.656 \times 10^4\ \frac{\text{lb}}{\text{ft}^2}}$$

Note that the above value is <u>not</u> the pressure at a stagnation point at the nose of a blunt body, because in slowing to zero velocity, the flow has to go through a shock wave, which is non-isentropic. The stagnation pressure at the nose of a body in a Mach 2 stream is the total pressure <u>behind a normal shock wave</u>, which is <u>lower</u> than the total pressure of the freestream, as calculated above. This stagnation pressure at the nose of a blunt body is given by Eq. (4.79).

$$\frac{p_{o_2}}{p_1} = \left[\frac{(\gamma+1)^2 M_\infty^2}{4\gamma M_2^2 - 2(\gamma-1)}\right]^{\frac{\gamma}{\gamma-1}} \left[\frac{1-\gamma+2\gamma M_1^2}{\gamma+1}\right] =$$

$$= \left[\frac{(2.4)^2 (2)^2}{4(1.4)(2)^2 - 2(0.4)}\right]^{1.4/0.4} \left[\frac{1-1.4+2(1.4)(2)^2}{2.4}\right] = 5.639$$

Hence,

$$p_{o_2} = 5.639\, p_\infty = 5.639\,(2116) = \boxed{1.193 \times 10^4\ \frac{\text{lb}}{\text{ft}^2}}$$

If Bernoulli's equation is used, the following <u>wrong</u> result for total pressure is obtained.

$$p_o = p_\infty + q_\infty = p_\infty + \frac{1}{2}\rho V_\infty^2 = p_\infty + \frac{\gamma}{2} p_\infty M_\infty^2$$

$$p_o = 2116 + 0.7\,(2116)\,(2)^2 = 0.804 \times 10^4\ \frac{\text{lb}}{\text{ft}^2}$$

Compared to the correct result of $1.656 \times 10^4\ \frac{\text{lb}}{\text{ft}^2}$, this leads to an error <u>51%</u>.

4.31 $\quad \dfrac{p_e}{p_o} = \left(1 + \dfrac{\gamma-1}{2} M_e^2\right)^{\frac{-\gamma}{\gamma-1}}$

$p_e = 5(1.01 \times 10^5)\,[1 + 0.2\,(3)^2]^{-3.5}$

$p_e = \boxed{1.37 \times 10^4\ \text{N/m}^2}$

$\dfrac{T_e}{T_o} = \left(1 + \dfrac{\gamma-1}{2} M_e^2\right)^{-1} = [1 + 0.2\,(3)^2]^{-1}$

$$T_e = (500)(0.357) = \boxed{178.6\text{K}}$$

$$\rho_e = \frac{p_e}{RT_e} = \frac{1.37 \times 10^4}{(287)(178.6)} = \boxed{0.267 \text{ kg/m}^3}$$

4.32 $\quad \dfrac{p_e}{p_o} = \left(1 + \dfrac{\gamma-1}{2}M_e^2\right)^{\frac{-\gamma}{\gamma-1}}$

Hence,

$$M_e^2 = \frac{2}{\gamma-1}\left[\left(\frac{p_e}{p_o}\right)^{\frac{1-\gamma}{\gamma}} - 1\right]$$

$M_e^2 = 5[(0.2)^{-0.286} - 1] = 2.92$

$M_e = 1.71$

$$\left(\frac{A_e}{A_t}\right)^2 = \frac{1}{M_e^2}\left[\frac{2}{\gamma+1}\left(1 + \frac{\gamma-1}{\gamma}\right)M^2\right]^{(\gamma+1)/(\gamma-1)}$$

$$\left(\frac{A_e}{A_t}\right)^2 = \frac{1}{(1.71)^2}[(0.833)(1 + 0.2(1.71)^2)]^6$$

$\dfrac{A_e}{A_t} = \boxed{1.35}$

4.33 $\quad \dfrac{q}{p_o} = \dfrac{\gamma}{2}\dfrac{p}{p_o}M^2 = \dfrac{\gamma}{2}M^2\left[1 + \dfrac{\gamma-1}{2}M^2\right]^{-\gamma/(\gamma-1)} = 0.7 M^2 (1 + 0.2 M^2)^{-3.5}$

M	M²	q/p₀
0	0	0
0.2	0.04	0.027
0.4	0.16	0.100
0.6	0.36	0.198
0.8	0.64	0.294
1.0	1.0	0.370
1.2	1.44	0.416
1.4	1.96	0.431
1.6	2.56	0.422
1.8	3.24	0.395
2.0	4.00	0.358

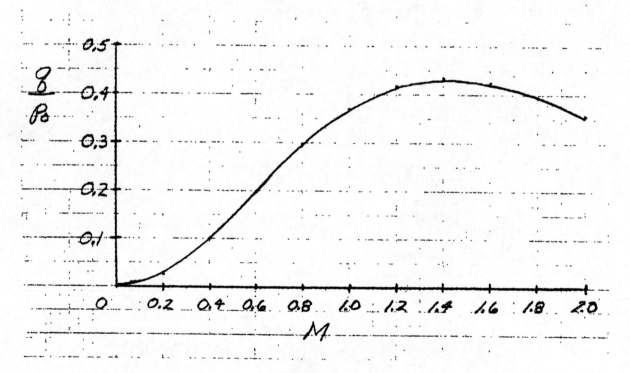

Note that the dynamic pressure increases with Mach number for M < 1.4 but decreases with Mach number for M > 1.4. I.e., in an isentropic nozzle expansion, there is a peak local dynamic pressure which occurs at M = 1.4

4.34 First, calculate the value of the Reynolds number.

$$Re_L = \frac{\rho_\infty V_\infty L}{\mu_\infty} = \frac{(1.225)(200)(3)}{(1.7894 \times 10^{-5})} = 4.10 \times 10^7$$

The dynamic pressure is

$$q_\infty = \frac{1}{2}\rho_\infty V_\infty^2 = \frac{1}{2}(1.225)(200)^2 = 2.45 \times 10^4 \text{ N/m}^2$$

Hence,

$$\delta_L = \frac{5.2L}{\sqrt{Re_L}} = \frac{5.2(3)}{\sqrt{4.1 \times 10^7}} = 0.0024\text{m} = \boxed{0.24 \text{ cm}}$$

and

$$C_f = \frac{1.328}{\sqrt{Re_L}} = \frac{1.328}{\sqrt{4.1 \times 10^7}} = 0.00021$$

The skin friction drag on one side of the plate is:

$$D_f = q_\infty S c_f = (2.45 \times 10^4)(3)(17.5)(0.00021)$$

$$D_f = 270\text{N}$$

The total skin friction drag, accounting for both the top and the bottom of the plate is twice this value, namely

Total $D_f = \boxed{540\text{N}}$

4.35 $\delta = \dfrac{0.37L}{(Re_L)^{0.2}} = \dfrac{0.37(3)}{(4.1 \times 10^7)^{0.2}} = 0.033\text{m} = \boxed{3.3 \text{ cm}}$

From problem 4.24, we find

$$\delta_{turbulent}/\delta_{laminar} = \frac{3.3}{0.24} = \boxed{13.75}$$

The turbulent boundary layer is more than an order of magnitude thicker than the laminar boundary layer.

$$C_f = \frac{0.074}{(Re_L)^{0.2}} = \frac{0.074}{(4.1 \times 10^7)^{0.2}} = 0.0022$$

The skin friction drag on one side is then

$$D_f = q_\infty S c_f = (2/45 \times 10^4)(3)(17.5)(0.0022)$$

$$D_f = 2830 N$$

The total, accounting for both top and bottom is

Total $D_f = \boxed{5660 N}$

From problem 4.24, we find

$$\left(D_{f_{turbulent}}\right)/\left(D_{f_{laminar}}\right) = \frac{5660}{540} = \boxed{10.5}$$

The turbulent skin friction drag is an order of magnitude larger than the laminar value.

4.36 $Re_{x_{cr}} = \dfrac{\rho_\infty V_\infty x_{cr}}{\mu_\infty}$

$$x_{cr} = Re_{x_{cr}} \left(\frac{\mu_\infty}{\rho_\infty V_\infty}\right) = \frac{(10^6)(1.789 \times 10^{-5})}{(1.225)(200)}$$

$$x_{cr} = 7.3 \times 10^{-2} m$$

```
                              ↓ V∞ = 200 m/sec
          ↓         ┌─────────────────────────┐
  7.3×10⁻² m        │   LAMINAR FLOW      A   │
          ↑         │ ─ ─ ─ ─ ─ ─ ─ ─ ─ ─ ─ ─ │
                    │                         │
                    │   TURBULENT FLOW    B   │
                    │                         │
                    └─────────────────────────┘
```

The turbulent drag that would exist over the first 7.3×10^{-2} m of chord length from the leading edge (area A) is

$$D_{f_A} = \frac{0.074}{(Re_{cr})^{0.2}} q_\infty S_A \quad \text{(on one side)}$$

$$D_{f_A} = \frac{0.074}{(10^6)^{0.2}} (2.45 \times 10^4)(7.3 \times 10^{-2})(17.5)$$

$$D_{f_A} = 146 \text{N} \quad \text{(on one side)}$$

From problem 4.25, the turbulent drag on one side, assuming both areas A and B to be turbulent, is 2830N. Hence, the turbulent drag on area B alone is:

$$D_{f_B} = 2830 - 146 - 2684 \text{N} \quad \text{(turbulent)}$$

The laminar drag on area A is

$$D_{f_A} = \frac{1.328}{(Re_{cr})^{0.5}} q_\infty S$$

$$D_{f_A} = \frac{1.328}{(10^6)^{0.5}} (2.45 \times 10^4)(7.3 \times 10^{-2})(17.5)$$

$$D_{f_A} = 42 \text{N} \quad \text{(laminar)}$$

Hence, the skin friction drag on one side, assuming area A to be laminar and area B to be turbulent is

$$D_f = D_{f_A} \text{ (laminar)} + D_{f_B} \text{ (turbulent)}$$

$$D_f = 42 + 2684 = 2726 \text{N}$$

The total drag, accounting for both sides, is

$$\text{Total } D_f = \boxed{5452 \text{N}}$$

Note: By comparing the results of this problem with those of problem 4.25, we see that the flow over the wing is mostly turbulent, which is usually the case for real airplanes in flight.

4.37 The relation between changes in pressure and velocity at a point in an inviscid flow is given by the Euler equation, Eq. (4.8)

$$dp = -\rho V dV$$

Letting s denote distance along the streamline through the point, Eq. (4.8) can be written as

$$\frac{dp}{ds} = -\rho V \frac{dV}{ds}$$

or, $$\frac{dp}{ds} = -\rho V^2 \frac{(dV/V)}{ds}$$

(a) $\frac{(dV/V)}{ds} = 0.02$ per millimeter

Hence,

$$\frac{dp}{ds} = -(1.1)(100)^2(0.02) = 220 \; \frac{N}{m^2} \text{ per millimeter}$$

(b) $$\frac{dp}{ds} = -(1.1)(1000)^2(0.02) = 22{,}000 \; \frac{N}{m^2} \text{ per millimeter}$$

Conclusion: At a point in a high-speed flow, it requires a much larger pressure gradient to achieve a given percentage change in velocity than for a low speed flow, everything else being equal.

4.38 We use the fact that total pressure is constant in an isentropic flow. From Eq. (4.74) applied in the freestream.

$$\frac{p_o}{p_\infty} = \left(1 + \frac{\gamma-1}{\gamma} M_\infty^2\right)^{\frac{\gamma}{\gamma-1}} = [1 + 0.2 (0.7)^2]^{3.5} = 1.387$$

From Eq. (4.74) applied at the point on the wing,

$$\frac{p_o}{p} = \left(1 + \frac{\gamma-1}{2} M^2\right)^{\frac{\gamma}{\gamma-1}} = [1 + 0.2 (1.1)^2]^{3.5} = 2.135$$

Hence,

$$p = \left[\left(\frac{p_o}{p_\infty}\right) / \left(\frac{p_o}{p}\right)\right] p_\infty = \left(\frac{1.387}{2.135}\right) p_\infty = 0.65 \; p_\infty$$

At a standard altitude of 3 km, from Appendix A, $p_\infty = 7.0121 \times 10^4$ N/m². Hence,

$$p = (0.65)(7.0121 \times 10^4) = \boxed{4.555 \times 10^4 \text{ N/m}^2}$$

4.39 This problem is simply asking what is the equivalent airspeed, as discussed in Section 4.12. Hence,

$$V_e = V \left(\frac{\rho}{\rho_s}\right)^{1/2} = (800) \left(\frac{1.0663 \times 10^{-3}}{2.3769 \times 10^{-3}}\right)^{1/2} = \boxed{535.8 \text{ ft.sec}}$$

4.40 (a) From Eq. (4.88)

$$\left(\frac{A_e}{A_t}\right)^2 = \frac{1}{M_e^2} \left[\frac{2}{\gamma+1}\left(1 + \frac{\gamma-1}{2} M_e^2\right)\right]^{\frac{\gamma+1}{\gamma-1}} = \frac{1}{(10)^2} \left\{\frac{2}{2.4}[1 + 0.2 \, (10)^2]\right\}^6 = 2.87 \times 10^5$$

Hence:

$$\frac{A_e}{A_t} = \sqrt{2.87 \times 10^5} = \boxed{535.9}$$

(b) From Eq. (4.87)

$$\frac{p_o}{p_e} = \left(1 + \frac{\gamma-1}{2} M_e^2\right)^{\frac{\gamma}{\gamma-1}} = [1 + 0.2 (10)^2]^{3.5} = 4.244 \times 10^4$$

At a standard altitude of 55 km, p = 48.373 N/m². Hence

$$p_o = (4.244 \times 10^4)(48.373) = 2.053 \times 10^6 \text{ N/m}^2 = \boxed{20.3 \text{ atm}}$$

(c) From Eq. (4.85)

$$\frac{T_o}{T_e} = 1 + \frac{\gamma-1}{2} M_e^2 = 1 + 0.2 (10)^2 = 21$$

At a standard altitude of 55 km, T = 275.78 K. Hence,

$$T_o = 275.78 (21) = \boxed{5791 \text{ K}}$$

Examining the above results, we note that:

1. The required expansion ratio of 535.9 is <u>huge</u>, but is readily manufactured.

2. The required reservoir pressure of 20.3 atm is large, but can be handled by proper design of the reservoir chamber.

3. The required reservoir temperature of 5791 K is tremendously large, especially when you remember that the surface temperature of the sun is about 6000 K. For a continuous flow hypersonic tunnel, such high reservoir tempertures can not be handled. In practice, a reservoir temperature of about half this value or less is employed, with the sacrifice made that "true temperature" simulation in the test stream is not obtained.

4.41 The speed of sound in the test stream is

$$a_e = \sqrt{\gamma \, R \, T_e} = \sqrt{(1.4)(287)(275.78)} = 332.9 \text{ m/sec}.$$

Hence,

$$V_e = M_e \, a_e = 10 \, (332.9) = \boxed{3329 \text{ m/sec}}$$

4.42 (a) From Eq. 4.88, for $M_e = 20$

$$\left(\frac{A_e}{A_t}\right)^2 = \frac{1}{M_e^2}\left[\frac{2}{\gamma+1}\left(1+\frac{\gamma-1}{2}M_e^2\right)\right]^{\frac{\gamma+1}{\gamma-1}} = \frac{1}{(20)^2}\left\{\frac{2}{2.4}\left[1+0.2\,(20)^2\right]\right\}^6 = 2.365 \times 10^8$$

Hence:

$$\frac{A_e}{A_t} = \boxed{15{,}377}$$

(b) From Eq. (4.85)

$$\frac{T_o}{T_e} = \left(1+\frac{\gamma-1}{2}M_e^2\right)^{-1} = [1+(0.2)(20)^2]^{-1} = 0.01235$$

Hence,

$$T_e = (5791)(0.01235) = 71.5 \text{ K}$$

$$a_e = \sqrt{\gamma \, R \, T_e} = \sqrt{(1.4)(287)(71.5)} = 169.5 \text{ m/sec}$$

$$V_e = M_e \, a_e = 20 \, (169.5) = \boxed{3390 \text{ m/sec}}$$

Comments:

1. To obtain Mach 20, i.e., to double the Mach number in this case, the expansion ratio must be increased by a factor of $15{,}377/535.9 = 28.7$. High hypersonic Mach numbers demand wind tunnels with very large exit-to-throat ratios. In practice, this is usually obtained by designing the nozzle with a <u>small throat area</u>.

2. Of particular interest is that the <u>exit velocity is increased by a very small amount</u>, namely by only 61 m/sec, although the exit Mach number has been doubled. The higher Mach number of 20 is achieved not by a large increase in exit velocity by rather by a <u>large decrease in the speed of sound at the exit</u>. This is characteristic of most conventional hypersonic wind tunnels -- the higher Mach numbers are not associated with corresponding increases in the test section flow velocities.

5.1 Assume the moment is governed by

$$M = f(V_\infty, \rho_\infty, S, \mu_\infty, a_\infty)$$

More specifically:

$$M = Z \, V_\infty^a \, \rho_\infty^b \, S^d \, a_\infty^e \, \mu_\infty^f$$

Equating the dimensions of mass, m, length, ℓ, and time t, and considering Z dimensionless,

$$\frac{m\ell^2}{t^2} = \left(\frac{\ell}{t}\right)^a \left(\frac{m}{\ell^3}\right)^b (\ell^2)^d \left(\frac{\ell}{t}\right)^e \left(\frac{m}{\ell t}\right)^f$$

$1 = b + f$ (For mass)

$2 = a - 3b + 2d + e - f$ (For length)

$-2 = -a - e - f$ (for time)

Solving a, b, and d in terms of e and f,

$b = 1 - f$

and, $a = 2 - e - f$

and, $2 = 2 - e - f - 3 + 3f + 2d + e - f$

or $0 = -3 + f + 2d$

$$d = \frac{3-f}{2}$$

Hence,

$$M = Z\, V_\infty^{2-e-f}\, \rho_\infty^{1-f}\, S^{(3-f)/2}\, a_\infty^e\, \mu_\infty^f$$

$$= Z\, V_\infty^2\, \rho_\infty\, S S^{1/2} \left(\frac{a_\infty}{V_\infty}\right)^e \left(\frac{\mu_\infty}{V_\infty \rho_\infty S^{1/2}}\right)^f$$

Note that $S^{1/2}$ is a characteristic length; denote it by the chord, c.

$$M = \rho_\infty\, V_\infty^2\, S\, c\, Z \left(\frac{a_\infty}{V_\infty}\right)^e \left(\frac{\mu_\infty}{V_\infty \rho_\infty c}\right)^f$$

However, $a_\infty/V_\infty = 1/M_\infty$

and $\dfrac{\mu_\infty}{V_\infty \rho_\infty c} = \dfrac{1}{Re}$

Let

$$Z \left(\frac{1}{M_\infty}\right)^e \left(\frac{1}{Re}\right)^f = \frac{c_m}{2}$$

where c_m is the moment coefficient. Then, as was to be derived, we have

$$M = \frac{1}{2} \rho_\infty\, V_\infty^2\, c\, c_m$$

or, $M = q_\infty\, S\, c\, c_m$

5.2 From Appendix D, at 5° angle of attack,

$$c_\ell = 0.67$$

$$c_{m_{c/4}} = -0.025$$

(Note: Two sets of lift and moment coefficient data are given for the NACA 1412 airfoil -- with and without flap deflection. Make certain to read the code properly, and use only the unflapped data, as given above. Also, note that the scale for $c_{m_{c/4}}$ is different than that for c_ℓ -- be careful in reading the data.)

With regard to c_d, first check the Reynolds number,

$$\mathrm{Re} = \frac{\rho_\infty V_\infty c}{\mu_\infty} = \frac{(0.002377)(100)(3)}{(3.7373 \times 10^{-7})}$$

$$\mathrm{Re} = 1.9 \times 10^6$$

In the airfoil data, the closest Re is 3×10^6. Use c_d for this value.

$$c_d = 0.007 \quad (\text{for } c_\ell = 0.67)$$

The dynamic pressure is

$$q_\infty = \frac{1}{2}\rho_\infty V_\infty^2 = \frac{1}{2}(0.002377)(100)^2 = 11.9 \text{ lb/ft}^2$$

The area per unit span is $S = 1(c) = (1)(3) = 3 \text{ ft}^2$

Hence, per unit span,

$$L = q_\infty S\, c_\ell = (11.9)(3)(0.67) = \boxed{23.9 \text{ lb}}$$

$$D = q_\infty S\, c_d = (11.9)(3)(0.007) = \boxed{0.25 \text{ lb}}$$

$$M_{c/4} = q_\infty S\, c\, c_{m_{c/4}} = (11.9)(3)(3)(-0.025) = \boxed{-2.68 \text{ ft.lb}}$$

5.3 $\rho_\infty = \dfrac{p_\infty}{RT_\infty} = \dfrac{(1.01 \times 10^5)}{(287)(303)} = 1.61$ kg/m³

From Appendix D,

$c_\ell = 0.98$

$c_{m_{c/4}} = -0.012$

Checking the Reynolds number, using the viscosity coefficient from the curve given in Chapter 4,

$\mu_\infty = 1.82 \times 10^{-5}$ kg/m sec at T = 303K,

$Re = \dfrac{\rho_\infty V_\infty c}{\mu_\infty} = \dfrac{(1.157)(42)(0.3)}{1.82 \times 10^{-5}} = 8 \times 10^5$

This Reynolds number is considerably less than the lowest value of 3×10^6 for which data is given for the NACA 23012 airfoil in Appendix D. Hence, we can use this data only to give an educated guess; use

$c_d \approx 0.01$, which is about 10 percent higher than the value of 0.009 given for Re = 3×10^6

The dynamic pressure is

$q_\infty = \dfrac{1}{2}\rho_\infty V_\infty^2 = \dfrac{1}{2}(1.161)(42)^2 = 1024$ N/m²

The area per unit span is S = (1)(0.3) = 0.3 m². Hence,

$L = q_\infty\, S\, c_\ell = (1024)(0.3)(0.98) = \boxed{301\text{N}}$

$D = q_\infty\, S\, c_d = (1024)(0.3)(0.01) = \boxed{3.07\text{N}}$

$M_{c/4} = q_\infty\, S\, c\, c_m = (1024)(0.3)(0.3)(-0.012) = \boxed{-1.1\text{Nm}}$

5.4 From the previous problem, $q_\infty = 1020$ N/m²

$$L = q_\infty S c_\ell$$

Hence,

$$c_\ell = \frac{L}{q_\infty S}$$

The wing area $S = (2)(0.3) = 0.6 \text{ m}^2$

Hence,

$$c_\ell = \frac{200}{(1024)(0.6)} = 0.33$$

From Appendix D, the angle of attack which corresponds to this lift coefficient is

$$\boxed{\alpha = 2°}$$

5.5 From Appendix D, at $\alpha = 4°$,

$$c_\ell = 0.4$$

Also, $V_\infty = 120 \left(\frac{88}{60}\right) = 176$ ft/sec

$$q_\infty = \frac{1}{2}\rho_\infty V_\infty^2 = \frac{1}{2}(0.002377)(176)^2 = 36.8 \text{ lb/ft}^2$$

$$L = q_\infty S c_\ell$$

$$S = \frac{L}{q_\infty c_\ell} = \frac{29.5}{(36.8)(0.4)} = \boxed{2 \text{ ft}^2}$$

5.6 $L = q_\infty S c_\ell$

$D = q_\infty S c_d$

Hence,

$$\frac{L}{D} = \frac{q_\infty S\, c_\ell}{q_\infty S\, c_d} = \frac{c_\ell}{c_d}$$

We must tabulate the values of c_ℓ/c_d for various angles of attack, and find where the maximum occurs. For example, from Appendix D, at $\alpha = 0°$,

$c_\ell = 0.25$

$c_d = 0.006$

Hence

$$\frac{L}{D} = \frac{c_\ell}{c_d} = \frac{0.25}{0.006} = 41.7$$

A tabulation follows.

α	0°	1°	2°	3°	4°	5°	6°	7°	8°	9°
c_ℓ	0.25	0.35	0.45	0.55	0.65	0.75	0.85	0.95	1.05	1.15
c_d	0.006	0.006	0.006	0.0065	0.0072	0.0075	0.008	0.0085	0.0095	0.0105
$\frac{c_\ell}{c_d}$	41.7	58.3	75	84.6	90.3	100	106	112	111	110

From the above tabulation,

$$\left(\frac{L}{D}\right)_{max} \approx \boxed{112}$$

5.7 At sea level

$\rho_\infty = 1.225 \text{ kg/m}^3$

$$p_\infty = 1.01 \times 10^5 \text{ N/m}^2$$

Hence,

$$q_\infty = \frac{1}{2}\rho_\infty V_\infty^2 = \frac{1}{2}(1.225)(50)^2 = 1531 \text{ N/m}^2$$

From the definition of pressure coefficient,

$$C_p = \frac{p - p_\infty}{q_\infty} = \frac{(0.95 - 1.01) \times 10^5}{1531} = \boxed{-3.91}$$

5.8 The speed is low enough that incompressible flow can be assumed. From Bernoulli's equation,

$$p + \frac{1}{2}\rho V_\infty^2 = p_\infty + \frac{1}{2}\rho_\infty V_\infty^2 = p_\infty + q_\infty$$

$$C_p = \frac{p - p_\infty}{q_\infty} = \frac{q_\infty - \frac{1}{2}\rho V^2}{q_\infty} = 1 - \frac{\frac{1}{2}\rho V^2}{\frac{1}{2}\rho_\infty V_\infty^2}$$

Since $\rho = \rho_\infty$ (constant density)

$$C_p = 1 - \left(\frac{V}{V_\infty}\right)^2 = 1 - \left(\frac{62}{55}\right)^2 = 1 - 1.27 = \boxed{-0.27}$$

5.9 The flow is low speed, hence assumed to be incompressible. From problem 5.8,

$$C_p = 1 - \left(\frac{V}{V_\infty}\right)^2 = 1 - \left(\frac{195}{160}\right)^2 = \boxed{-0.485}$$

5.10 The speed of sound is

$$a_\infty = \sqrt{\gamma RT_\infty} = \sqrt{(1.4)(1716)(510)} = 1107 \text{ ft/sec}$$

Hence,

$$M_\infty = \frac{V_\infty}{a_\infty} = \frac{700}{1107} = 0.63$$

In problem 5.9, the pressure coefficient at the given point was calculated as -0.485. However, the conditions of problem 5.9 were low speed, hence we identify

$$C_{p_o} = 0.485$$

At the new, higher free stream velocity, the pressure coefficient must be corrected for compressibility. Using the Prandtl-Glauert Rule, the high speed pressure coefficient is

$$C_p = \frac{C_{p_o}}{\sqrt{1-M_\infty^2}} = \frac{-0.485}{\sqrt{1-(0.63)^2}} = \boxed{-0.625}$$

5.11 The formula derived in problem 5.8, namely

$$C_p = 1 - \left(\frac{V}{V_\infty}\right)^2,$$

utilized Bernoulli's equation in the derivation. Hence, it is <u>not valid</u> for compressible flow. In the present problem, check the Mach number.

$$a_\infty = \sqrt{\gamma RT_\infty} = \sqrt{(1.4)(1716)(505)} = 1101 \text{ ft/sec}$$

$$M_\infty = \frac{780}{1101} = 0.708.$$

The flow is clearly compressible! To obtain the pressure coefficient, first calculate ρ_∞ from the equation of state.

$$\rho_\infty = \frac{p_\infty}{RT_\infty} = \frac{2116}{(1716)(505)} = 0.00244 \text{ slug/ft}^3$$

To find the pressure at the point on the wing where V = 850 ft/sec, first find the temperature from the energy equation

$$c_p T + \frac{V^2}{V} = c_p T_\infty + \frac{V_\infty^2}{2}$$

$$T = T_\infty + \frac{V_\infty^2 - V^2}{2c_p}$$

The specific heat at constant pressure for air is

$$c_p = \frac{\gamma R}{\gamma - 1} = \frac{(1.4)(1716)}{(1.4 - 1)} = 6006 \frac{\text{ft lb}}{\text{slug R}}$$

Hence,

$$T = 505 + \frac{780^2 - 850^2}{2(6006)} = 505 - 9.5 = 495.5 R$$

Assuming isentropic flow

$$\frac{p}{p_\infty} = \left(\frac{T}{T_\infty}\right)^{\frac{\gamma}{\gamma-1}}$$

$$p = (2116)\left(\frac{495.5}{505}\right)^{3.5} = 1980 \text{ lb/ft}^2$$

From the definition of C_p

$$C_p = \frac{p - p_\infty}{q_\infty} = \frac{p - p_\infty}{\frac{1}{2}\rho_\infty V_\infty^2} = \frac{1980 - 2116}{\frac{1}{2}(0.00244)(780)^2}$$

$$C_p = \boxed{-0.183}$$

5.12 A velocity of 100 ft/sec is low speed. Hence, the desired pressure coefficient is a low speed value, C_{p_o}.

$$C_p = \frac{C_{p_o}}{\sqrt{1-M_\infty^2}}$$

From problem 5.11,

C_p = -0.183 and M_∞ = 0.708. Thus, $0.183 = \dfrac{C_{p_o}}{\sqrt{1-(0.708)^2}}$

$C_{p_o} = (-0.183)(0.706) = \boxed{-0.129}$

5.13 Recall that the airfoil data in Appendix D is for low speeds. Hence, at $\alpha = 4°$, c_{ℓ_o} = 0.58.

Thus, from the Prandtl-Glauert rule,

$$c_\ell = \frac{c_{\ell_o}}{\sqrt{1-M_\infty^2}} = \frac{0.58}{\sqrt{1-(0.8)^2}} = \boxed{0.97}$$

5.14 The lift coefficient measured is the high speed value, c_ℓ. Its low speed counterpart is c_{ℓ_o}, where

$$c_\ell = \frac{c_{\ell_o}}{\sqrt{1-M_\infty^2}}$$

Hence,

$$c_{\ell_o} = (0.85)\sqrt{1-(0.7)^2} = 0.607$$

For this value, the low speed data in Appendix D yield

$$\boxed{\alpha = 2°}$$

5.15 First, obtain a curve of $C_{p,cr}$ versus M_∞ from

$$C_{p,cr} = \frac{2}{\gamma M_\infty^2}\left[\left(\frac{2+(\gamma-1)M_\infty^2}{\gamma+1}\right)^{\gamma/(\gamma-1)} - 1\right]$$

Some values are tabulated below for $\gamma = 1.4$.

M_∞	0.4	0.5	0.6	0.7	0.8	0.9	1.0
$C_{p,cr}$	-3.66	-2.13	-1.29	-0.779	-0.435	-0.188	0

Now, obtain the variation of the minimum pressure coefficient, C_p, with M_∞, where $C_{p_o} = -0.90$. From the Prandtl-Glauert rule,

$$C_p = \frac{C_{p_o}}{\sqrt{1-M_\infty^2}}$$

$$C_p = \frac{-0.90}{\sqrt{1-M_\infty^2}}$$

Some tabulated values are:

M_∞	0.4	0.5	0.6	0.7	0.8	0.9
C_p	-0.98	-1.04	-1.125	-1.26	-1.5	-2.06

A plot of the two curves is given on the next page.

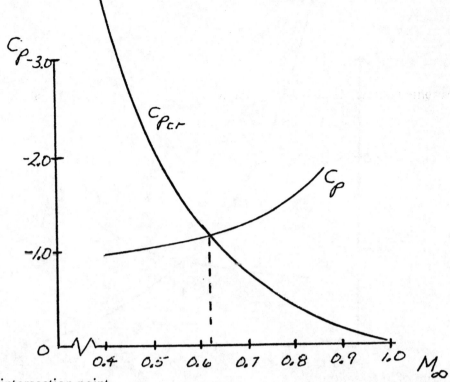

From the intersection point,

$M_{cr} = \boxed{0.62}$

5.16 The curve of $C_{p,cr}$ versus M_∞ has already been obtained in the previous problem; it is a universal curve, and hence can be used for this and all other problems. We simply have to obtain the variation of C_p with M_∞ from the Prandtl-Glauert rule, as follows:

$$C_p = \frac{C_{p_o}}{\sqrt{1-M_\infty^2}} = \frac{-0.65}{\sqrt{1-M_\infty^2}}$$

M_∞	0.4	0.5	0.6	0.7	0.8	0.9
C_p	-0.71	0.75	-0.81	-0.91	-1.08	-1.49

The results are plotted below.

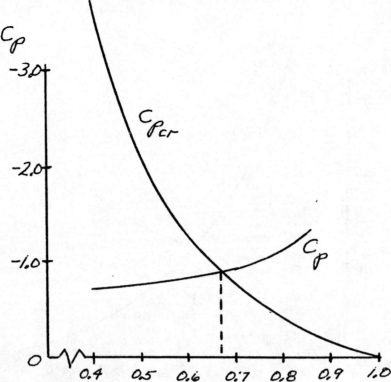

From the point of intersection,

$M_{cr} = \boxed{0.68}$

Please note that, comparing problems 5.15 and 5.16, the critical Mach number for a given airfoil is somewhat dependent on angle of attack for the simple reason that the value of the minimum pressure coefficient is a function of angle of attack. When a critical Mach number is stated for a given airfoil in the literature, it is usually for a small (cruising) angle of attack.

5.17 Mach angle = μ = arc sin (1/M)

μ = arc sin (1/2) = $\boxed{30°}$

5.18

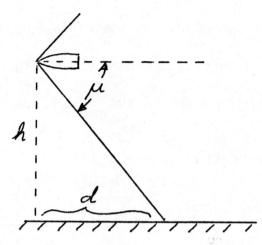

$$\mu = \sin^{-1}\left(\frac{1}{M}\right) = \sin^{-1}\left(\frac{1}{2.5}\right) = 23.6°$$

$$d = h/\tan\mu = \frac{10 \text{ km}}{0.436} = \boxed{22.9 \text{ km}}$$

5.19 At 36,000 ft, from Appendix B,

$T_\infty = 390.5°R$

$\rho_\infty = 7.1 \times 10^{-4}$ slug/ft^3

Hence,

$$a_\infty = \sqrt{\gamma R T_\infty} = \sqrt{(1.4)(1716)(390.5)} = 969 \text{ ft/sec}$$

$$V_\infty = a_\infty M_\infty = (969)(2.2) = 2132 \text{ ft/sec}$$

$$q_\infty = \frac{1}{2}\rho_\infty V_\infty^2 = \frac{1}{2}(7.1 \times 10^{-4})(2132)^2 = 1614 \text{ lb/ft}^2$$

In level flight, the airplane's lift must balance its weight, hence

$L = W = 16,000$ lb.

From the definition of lift coefficient,

$$C_L = L/q_\infty S = 16,000/(1614)(210) = 0.047$$

Assume that all the lift is derived from the wings (this is not really true because the fuselage and horizontal tail also contribute to the airplane lift.) Moreover, assume the wings can be approximated by a thin flat plate. Hence, the lift coefficient is given approximately by

$$C_L = \frac{4\alpha}{\sqrt{M_\infty^2 - 1}}$$

Solve for α,

$$\alpha = \frac{1}{4} C_L \sqrt{M_\infty^2 - 1} = \frac{1}{4}(0.047)\sqrt{(2.2)^2 - 1}$$

$\alpha = 0.023$ radians (or 1.2 degrees)

The wave drag coefficient is approximated by

$$C_{D_w} = \frac{4\alpha}{\sqrt{M_\infty^2 - 1}} = \frac{4(0.023)^2}{\sqrt{(2.2)^2 - 1}} = 0.00108$$

Hence,

$$D_w = q_\infty \, S \, C_{D_w} = (1614)(210)(0.00108)$$

$D_w = \boxed{366 \text{ lb}}$

5.20 (a) At 50,000 ft, $\rho_\infty = 3.6391 \times 10^{-4}$ slug/ft^3 and $T_\infty = 390°R$. Hence,

$$a_\infty = \sqrt{\gamma R T_\infty} = \sqrt{(1.4)(1716)(390)} = 968 \text{ ft/sec}$$

and $V_\infty = a_\infty M_\infty = (968)(2.2) = 2130$ ft/sec

The viscosity coefficient at $T_\infty = 390°R = 216.7K$ can be estimated from an extrapolation of the straight line given in Fig. 4.30. The slope of this line is

$$\frac{d\mu}{dT} = \frac{(2.12-1.54) \times 10^{-5}}{(350-250)} = 5.8 \times 10^{-8} \frac{kg}{(m)(sec)(K)}$$

Extrapolating from the sea level value of $\mu = 1.7894 \times 10^{-5}$ kg/(m)(sec), we have at $T_\infty = 216.7$ K.

$$\mu_\infty = 1.7894 \times 10^{-5} - (5.8 \times 10^{-8})(288 - 216.7)$$

$$\mu_\infty = 1.37 \times 10^{-5} \text{ kg/(m)(sec)}$$

Converting to english engineering units, using the information in Chapter 4, we have

$$\mu_\infty = \frac{1.37 \times 10^{-5}}{1.7894 \times 10^{-5}} (3.7373 \times 10^{-7} \frac{slug}{ft\ sec}) = 2.86 \times 10^{-7} \frac{slug}{ft\ sec}$$

Finally, we can calculate the Reynolds number for the flat plate:

$$Re_L = \frac{\rho_\infty V_\infty L}{\mu_\infty} = \frac{3.6391 \times 10^{-4}(2130)(202)}{2.86 \times 10^{-7}} = 5.47 \times 10^8$$

Thus, from Eq. (4.100) reduced by 20 percent

$$C_f = (0.8)\frac{0.074}{(Re_L)^{0.2}} = (0.8)\frac{0.074}{(5.74 \times 10^8)^{0.2}} = 0.00106$$

The wave drag coefficient is estimated from

$$c_{d,w} = \frac{4\alpha^2}{\sqrt{M_\infty^2 - 1}}$$

where $\alpha = \frac{2}{57.3} = 0.035$ rad.

Thus,

$$c_{d,w} = \frac{4(0.035)^2}{\sqrt{(2.2)^2 - 1}} = 0.0025$$

Total drag coefficient = 0.0025 + (2)(0.00106) = $\boxed{0.00462}$

Note: In the above, C_f is multiplied by two, because Eq. (4.100) applied to only one side of the flat plate. In flight, both the top and bottom of the plate will experience skin friction, hence that total skin friction coefficient is 2(0.00106) = 0.00212.

(b) If α is increased to 5 degrees, where α = 5/57.3 - 0.00873 rad, then

$$c_{d,w} = \frac{4(0.0873)^2}{\sqrt{(2.2)^2 - 1}} = 0.01556$$

Total drag coefficient = 0.01556 + 2(0.00106) = $\boxed{0.0177}$

(c) In case (a) where the angle of attack is 2 degrees, the wave drag coefficient (0.0025) and the skin friction drag coefficient acting on both sides of the plate (2 x 0.00106 = 0.00212) are about the same. However, in case (b) where the angle of attack is higher, the wave drag coefficient (0.0177) is about eight times the total skin friction coefficient. This is because, as α increases, the strength of the leading edge shock increases rapidly. In this case, wave drag dominates the overall drag for the plate.

5.21 V_∞ = 251 km/h = $\left(251\frac{km}{h}\right)\left(\frac{1h}{3600\,sec}\right)\left(\frac{1000m}{1km}\right)$ = 69.7 m/sec

ρ_∞ = 1.225 kg/m³

$q_\infty = \frac{1}{2}\rho_\infty V_\infty^2 = \frac{1}{2}(1.225)(69.7)^2 = 2976$ N/m²

$C_L = \frac{L}{q_\infty S} = \frac{9800}{(2976)(16.2)} = 0.203$

$C_{D_i} = \frac{C_L^2}{\pi e AR} = \frac{(0.203)^2}{\pi(0.62)(7.31)} = 0.002894$

$$D_i = q_\infty \, S \, C_{D_i} = (2976)(16.2)(0.002894) = \boxed{139.5 \text{ N}}$$

5.22 $V_\infty = 85.5$ km/h $= 23.75$ m/sec

$$q_\infty = \frac{1}{2}\rho_\infty V_\infty^2 = \frac{1}{2}(1.225)(23.75)^2 = 345 \text{ N/m}^2$$

$$C_L = \frac{L}{q_\infty S} = \frac{9800}{(345)(16.2)} = 1.75$$

$$C_{D_i} = \frac{C_L^2}{\pi e AR} = \frac{(1.75)^2}{\pi(0.62)(7.31)} = 0.215$$

$$D_i = q_\infty \, S \, C_{D_i} = (345)(16.2)(0.215) = \boxed{1202 \text{ N}}$$

Note: The induced drag at low speeds, such as near stalling velocity, is considerably larger than at high speeds, near maximum velocity. Compare the results of problems 5.20 and 5.21.

5.23 First, obtain the infinite wing lift slope. From Appendix D for a NACA 65-210 airfoil,

$$C_\ell = 1.05 \text{ at } \alpha = 8°$$

$$C_\ell = 0 \text{ at } \alpha_{L=0} = -1.5°$$

Hence,

$$a_o = \frac{1.05 - 0}{8 - (-1.5)} = 0.11 \text{ per degree}$$

The lift slope for the finite wing is

$$a = \frac{a_o}{1 + \frac{57.3\, a_o}{\pi\, e_1 AR}} = \frac{0.11}{1 + \frac{57.3(0.11)}{\pi\, (.9)(5)}} = 0.076 \text{ per degree}$$

At $\alpha = 6°$,

$$C_L = a(\alpha - \alpha_{L=0}) = (0.076)[6 - (-1.5)] = \boxed{0.57}$$

The total drag coefficient is

$$C_D = c_d + \frac{C_L^2}{\pi e AR} = (0.004) + \frac{(0.57)^2}{\pi\, (0.9)(5)}$$

$$C_D = 0.004 + 0.023 = \boxed{0.027}$$

5.24 $\quad q_\infty = \frac{1}{2}\rho_\infty V_\infty^2 = \frac{1}{2}(0.002377)(100)^2 = 11.9 \text{ lb/ft}^2$

at $\alpha = 10°$, $L = 17.9$ lb. Hence

$$C_L = \frac{L}{q_\infty S} = \frac{17.9}{(11.9)(1.5)} = 1.0$$

at $\alpha = -2°$, $L = 0$. Hence $\alpha_{L=0} = -2°$

$$a = \frac{dC_L}{d\alpha} = \frac{1.0 - 0}{[10 - (-2)]} = 0.083 \text{ per degree}$$

This is the <u>finite</u> wing lift slope.

$$a = \frac{a_o}{1 + \frac{57.3\, a_o}{\pi\, eAR}}$$

Solve for a_o.

$$a_o = \frac{a}{1 - \frac{57.3\ a}{\pi\ eAR}} = \frac{0.083}{1 + \frac{57.3(0.083)}{\pi\ (0.95)(6)}}$$

$a_o = \boxed{0.11 \text{ per degree}}$

5.25 At $\alpha = -1°$, the lift is zero. Hence, the total drag is simply the profile drag.

$$C_D = c_d + \frac{C_L^2}{\pi eAR} = c_d + 0 = c_d$$

$$q_\infty = \frac{1}{2}\rho_\infty V_\infty^2 = \frac{1}{2}(0.002377)(130)^2 = 20.1 \text{ lb/ft}^2$$

Thus, at $\alpha = \alpha_{L=0} = -1°$

$$c_d = \frac{D}{q_\infty S} = \frac{0.181}{(20.1)(1.5)} = 0.006$$

At $\alpha = 2°$, assume that c_d has not materially changed, i.e., the "drag bucket" of the profile drag curve (see Appendix D) extends at least from $-1°$ to $2°$, where c_d is essentially constant. Thus, at $\alpha = 2°$,

$$C_L = \frac{L}{q_\infty S} = \frac{5}{(20.1)(1.5)} = 0.166$$

$$C_D = \frac{D}{q_\infty S} = \frac{0.23}{(20.1)(1.5)} = 0.00763$$

However:

$$C_D = c_d + \frac{C_L^2}{\pi eAR}$$

$$0.00763 = 0.006 + \frac{(0.166)^2}{\pi\ e(6)} = 0.006 + \frac{0.00146}{e}$$

$$e = \boxed{0.90}$$

To obtain the lift slope of the airfoil (infinite wing), first calculate the finite wing lift slope.

$$a = \frac{(0.166 - 0)}{[2 - (-2)]} = 0.055 \text{ per degree}$$

$$a_o = \frac{a}{1 - \dfrac{57.3\ a}{\pi\ eAR}} = \frac{0.055}{1 - \dfrac{57.3(0055)}{\pi\ (0.9)(6)}}$$

$$a_o = \boxed{0.068 \text{ per degree}}$$

5.26 $V_{stall} = \sqrt{\dfrac{2W}{\rho_\infty S\ C_{L_{max}}}} = \sqrt{\dfrac{2(7780)}{(1.225)(16.6)(2.1)}}$

$V_{stall} = \boxed{19.1 \text{ m/sec} = 68.7 \text{ km/h}}$

5.27 (a) $\alpha = \dfrac{5}{57.3} = 0.087$ radians

$c_\ell = 2\pi\alpha = 2\pi(0.087) = \boxed{0.548}$

(b) Using the Prandtl-Glauert rule,

$$c_\ell = \frac{c_{\ell_o}}{\sqrt{1 - M_\infty^2}} = \frac{0.548}{\sqrt{1 - (0.7)^2}} = \boxed{0.767}$$

(c) From Eq. (5.50)

$$c_\ell = \frac{4\alpha}{\sqrt{M_\infty^2 - 1}} = \frac{4(0.087)}{\sqrt{(2)^2 - 1}} = \boxed{0.2}$$

5.28 For $V_\infty = 21.8$ ft/sec at sea level

$$q_\infty = \frac{1}{2}\rho_\infty V_\infty^2 = \frac{1}{2}(0.002377)(21.8)^2 = 0.565 \text{ lb/ft}^2$$

1 ounce = 1/16 lb = 0.0625 lb.

$$C_L = \frac{L}{q_\infty S} = \frac{0.0625}{(0.565)(1)} = \boxed{0.11}$$

For a flat plate airfoil

$$c_\ell = 2\pi\alpha = 2\pi(3/57.3) = \boxed{0.329}$$

The difference between the higher value predicted by thin airfoil theory and the lower value measured by Cayley is due to the low aspect ratio of Cayley's test wing, and viscous effects at low Reynolds number.

5.29 From Eqs. (5.1) and (5.2), writtten in coefficient form

$$C_L = C_N \cos\alpha - C_A \sin\alpha$$

$$C_D = C_N \sin\alpha + C_A \cos\alpha$$

Hence:

$$C_L = 0.8 \cos 6° - 0.06 \sin 6° = 0.7956 - 0.00627 = \boxed{0.789}$$

$$C_D = 0.8 \sin\alpha + 0.06 \cos\alpha = 0.0836 + 0.0597 = \boxed{0.1433}$$

Note: At the relatively small angles of attack associated with normal airplane flight, C_L and C_N are essentially the same value, as shown in this example.

5.30 First solve for the angle of attack and the profile drag coefficient, which stay the same in this problem.

$$c_L = a\alpha = \frac{a_o \alpha}{1 + 57.3 \, a_o / (\pi \, e_1 AR)}$$

or, $\quad \alpha = \frac{C_L}{a_o}[1 + 57.3 \, a_o/\pi \, e_1 \, AR)]$

$$= \frac{0.35}{0.11}\{1 + 57.3 \, (0.11)/[\pi \, (0.9)(7)]\} = 4.2°$$

The profile drag can be obtained as follows.

$$C_D = \frac{C_L}{(C_L/C_D)} = \frac{0.35}{29} = 0.012$$

$$C_D = c_d + \frac{C_L^2}{\pi e AR}$$

or, $\quad c_d = C_D - \frac{C_L^2}{\pi e AR} = 0.012 - \frac{(0.35)^2}{\pi \, (.9)(7)} = 0.0062$

Increasing the aspect ratio at the same angle of attack increases C_L and reduces C_D. For AR = 10, we have

$$C_L = a\alpha = \frac{a_o \alpha}{1 + 57.3 \, a_o / (\pi \, e_1 AR)}$$

$$= \frac{(0.11)(4.2)}{1 + 57.3 \, (0.11)/[\pi \, (0.9)(10)]} = 0.3778$$

$$C_D = c_d + \frac{C_L^2}{\pi e AR} = 0.062 - \frac{(0.3778)^2}{\pi \, (.9)(10)} = 0.0062 + 0.005048 = 0.112$$

Hence, the new value of L/D is

$$\frac{C_L}{C_D} = \frac{0.3778}{0.0112} = \boxed{33.7}$$

6.1 (a) $V_\infty = 350$ km/hr $= 97.2$ m/sec

$$q_\infty = \frac{1}{2}\rho_\infty V_\infty^2 = \frac{1}{2}(1.225)(97.2)^2 = 5787 \text{ N/m}^2$$

$$C_L = \frac{W}{q_\infty S} = \frac{38220}{(5787)(27.3)} = 0.242$$

$$C_D = C_{D_o} + \frac{C_L^2}{\pi e AR} = 0.03 + \frac{(0.242)^2}{\pi(0.9)(7.5)}$$

$C_D = 0.03 + 0.0028 = 0.0328$

$C_L/C_D = 0.242/0.0328 = 7.38$

$$T_R = \frac{W}{C_L/C_D} = \frac{38220}{7.38} = \boxed{5179 \text{ N}}$$

(b) $q_\infty = \frac{1}{2}\rho_\infty V_\infty^2 = \frac{1}{2}(0.777)(97.2)^2 = 3670 \text{ N/m}^2$

$$C_L = \frac{W}{q_\infty S} = \frac{38220}{(3670)(27.3)} = 0.38$$

$$C_d = C_{D_o} + \frac{C_L^2}{\pi e AR} = 0.03 + \frac{(0.38)^2}{\pi(0.9)(7.5)}$$

$C_D = 0.03 + 0.0068 = 0.0368$

$C_L/C_D = 0.38/0.0368 = 10.3$

$$T_R = \frac{W}{C_L/C_D} = \frac{38220}{10.3} = \boxed{3711 \text{ N}}$$

6.2 $V_\infty = 200\,\dfrac{88}{60} = 293.3$ ft/sec

$$q_\infty = \frac{1}{2}\rho_\infty V_\infty^2 = \frac{1}{2}(0.002377)(293.3)^2 = 102.2 \text{ lb/ft}^2$$

$$C_L = \frac{L}{q_\infty S} = \frac{W}{q_\infty S} = \frac{5000}{(102.2)(200)} = 0.245$$

$$C_{D_i} = \frac{C_L^2}{\pi e AR} = \frac{(0.245)}{\pi (0.93)(8.5)} = 0.0024$$

Since the airplane is flying at the condition of maximum L/D, hence minimum thrust required, $C_{D_i} = C_{D_e}$. Thus,

$$C_D = C_{D_0} + C_{D_i} = 2\ C_{D_i} = 2(0.0024) = 0.1048$$

$$D = q_\infty S\ C_D = (102.2)(200)(0.0048) = \boxed{98.1 \text{ lb.}}$$

6.3 (a) Choose a velocity, say $V_\infty = 100$ m/sec

$$q_\infty = \frac{1}{2}\rho_\infty V_\infty^2 = \frac{1}{2}(1.225)(100)^2 = 6125 \text{ N/m}^2$$

$$C_L = \frac{W}{q_\infty S} = \frac{103047}{(6125)(47)} = 0.358$$

$$C_D = C_{D_0} + \frac{C_L^2}{\pi e AR} = 0.032 + \frac{(0.358)^2}{\pi (0.87)(6.5)}$$

$$C_D = 0.032 + 0.007 = 0.0392$$

$$T_R = \frac{W}{C_L/C_D} = \frac{103047}{9.13} = 11287 \text{ N}$$

$$P_R = T_R V_\infty = (11287)(100) = 1.129 \times 10^6 \text{ watts}$$

$$P_R = 1129 \text{ kw}$$

A tabulation for other velocities follows on the next page:

V (m/sec)	C_L	C_D	C_L/C_D	P_R (kw)
100	0.358	0.0392	9.13	1129
130	0.212	0.0345	6.14	2182
160	0.140	0.0331	4.23	3898
190	0.099	0.0325	3.05	6419
220	0.074	0.0323	2.29	9900
250	0.057	0.0322	1.77	14550
280	0.046	0.0321	1.43	20180
310	0.037	0.0321	1.15	27780

(b) $P_A = T_A V_\infty = (2)(40298) V_\infty = 80596 V_\infty$

The power required and power available curves are plotted below.

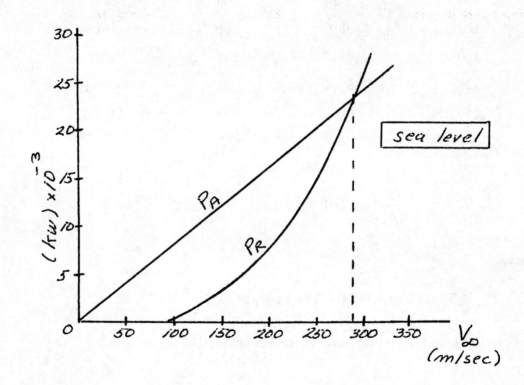

From the intersection of the P_A and P_R curves, we find,

$$V_{max} = \boxed{295 \text{ m/sec}} \text{ at sea level}$$

(c) At 5 km standard altitude, $\rho = 0.7364$ kg/m^3

Hence,

$$(\rho_o/\rho)^{1/2} = (1.225/0.7364)^{1/2} = 1.29$$

$$V_{alt} = (\rho_o/\rho)^{1/2} V_o = 1.29 \, V_o$$

$$P_{R \, alt} = (\rho_o/\rho)^{1/2} P_{R_o} = 1.29 \, P_{R_o}$$

From the results from part (a) above,

V_o (m/sec)	P_{R_o} (kw)	V_{alt} (m/sec)	$P_{R_{alt}}$ (kw)
100	1129	129	1456
130	2182	168	2815
160	3898	206	5028
190	6419	245	8281
220	9900	284	12771
250	14550	323	18770

(d) $T_{A_{alt}} = T_{A_o} \left(\dfrac{\rho}{\rho_o} \right) = 0.601 \, T_{A_o}$

Hence,

$$P_{A_{alt}} = T_{A_{alt}} \, V_\infty = (0.601)(80596) \, V_\infty = 48438 \, V_\infty$$

The power required and power available curves are plotted below.

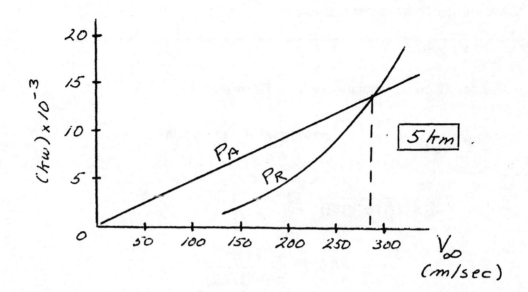

From the intersection of the P_R and P_A curves, we find

$$V_{max} = \boxed{290 \text{ m/sec}} \text{ at 5 km}$$

Comment: The mach numbers corresponding to the maximum velocities in parts(b) and (d) are as follows:

At sea level

$$a_\infty = \sqrt{\gamma R T_\infty} = \sqrt{(1.4)(287)(288.16)} = 340 \text{ m/sec}$$

$$M_\infty = \frac{V_\infty}{a_\infty} = \frac{295}{340} = 0.868$$

At 5 km altitude

$$a_\infty = \sqrt{\gamma R T_\infty} = \sqrt{(1.4)(287)(255.69)} = 321 \text{ m/sec}$$

$$M_\infty = \frac{V_\infty}{a_\infty} = \frac{290}{321} = 0.90$$

These mach numbers are slightly larger than what might be the actual drag-divergence Mach number for an airplane such as the A-10. Our calculations have not taken the large drag rise at drag-divergence into account. Hence, the maximum velocities calculated above are somewhat higher than reality.

6.4 (a) Choose a velocity, say $V_\infty = 100$ ft/sec

$$q_\infty = \frac{1}{2}\rho_\infty V_\infty^2 = \frac{1}{2}(0.002377)(100)^2 = 11.89 \text{ lb/ft}^2$$

$$C_L = \frac{W}{q_\infty S} = \frac{3000}{(11.89)(181)} = 1.39$$

$$C_D = C_{D_o} + \frac{C_L^2}{\pi e AR} = 0.027 + \frac{(1.39)^2}{\pi (0.91)(6.2)}$$

$$C_D = 0.027 + 0.109 = 0.136$$

$$T_R = \frac{W}{C_L/C_D} = \frac{3000}{(1.39)/(0.136)} = \frac{3000}{10.22} = 293.5 \text{ lb}$$

$$P_R = T_R V_\infty = (293.5)(100) = 29350 \text{ ft lb/sec}$$

In terms of horsepower,

$$P_R = \frac{29350}{550} = 53.4 \text{ hp}$$

A tabulation for other velocities follows:

V_∞ (ft/sec)	C_L	C_D	C_L/C_D	P_R (hp)
70	2.85	0.485	5.88	64.9
100	1.39	0.136	10.22	53.4
150	0.62	0.0487	12.73	64.3
200	0.349	0.0339	10.29	106
250	0.223	0.0298	7.48	182
300	0.155	0.0284	5.46	300
350	0.114	0.0277	4.12	463

(b) At sea level, maximum $P_A = 0.83(345) = 286$ hp. The power required and power available are plotted below.

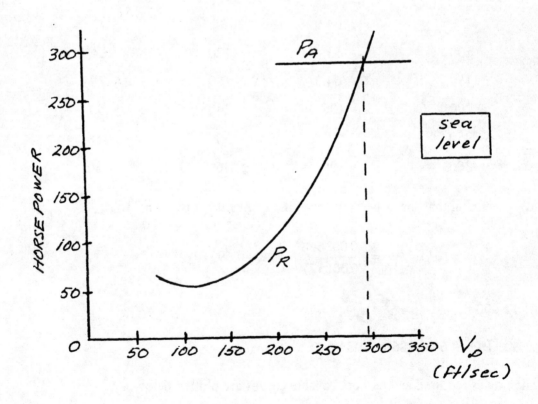

From the intersection of the P_A and P_R curves,

$$V_{max} = \boxed{295 \text{ ft/sec} = 201 \text{ mph}} \text{ at sea level}$$

(c) At a standard altitude of 12,000 ft,

$$\rho = 1.648 \times 10^{-3} \text{ slug/ft}^3$$

Hence,

$$(\rho_o/\rho)^{1/2} = (0.002377/0.001648)^{1/2} = 1.2$$

$$V_{alt} = (\rho_o/\rho)^{1/2} V_o = 1.2 \, V_o$$

$$P_{R_{alt}} = (\rho_o/\rho)^{1/2} P_{R_o} = 1.2 \, P_{R_o}$$

Using the results from part (a) above,

V_o (ft / sec)	P_{R_o} (hp)	V_{alt} (ft / sec)	$P_{R_{alt}}$ (hp)
70	64.9	84	77.9
100	53.4	120	64.1
150	64.3	180	76.9
200	106	240	127
250	182	300	218
300	300	360	360

(d) Assuming that the power output of the engine is proportional to ρ_∞,

$$P_{A_{alt}} = (\rho/\rho_o) P_{A_o} = \left(\frac{0.001648}{0.002377}\right) P_{A_o} = 0.693 \, P_{A_o}$$

At 12,000 ft,

$$P_A = 0.693 \, (286) = 198 \text{ hp}$$

The power required and power available curves are plotted below.

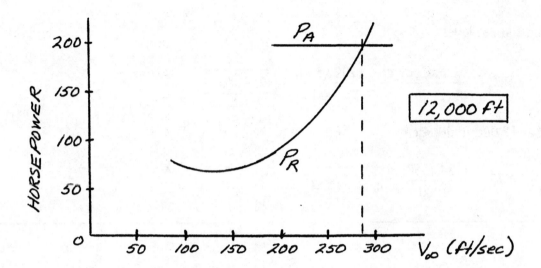

From the intersection of the P_A and P_R curves,

$$V_{max} = \boxed{290 \text{ ft/sec} = 198 \text{ mph}} \text{ at } 12{,}000 \text{ ft.}$$

6.5 From the P_A and P_B curves generated in problem 6.3, we find approximately:

excess power = 9000 kw at sea level

excess power = 5000 kw at 5 km

Hence, at sea level

$$R/C = \frac{\text{excess power}}{W} = \frac{9 \times 10^6 \text{ watts}}{1.0307 \times 10^5 \text{ N}} = \boxed{87.3 \text{ m/sec}}$$

and at 5 km altitude,

$$R/C = \frac{5 \times 10^6}{1.03047 \times 10^5} = \boxed{48.5 \text{ m/sec}}$$

6.6 From the P_A and P_R curves generated in problem 6.4, we find approximately:

excess power = 232 hp at sea level

excess power = 134 hp at 12,000 ft

Hence, at sea level,

$$R/C = \frac{\text{excess power}}{W} = \frac{(232)(550)}{3000} = \boxed{42.5 \text{ ft/sec}}$$

and at 12,000 ft altitude,

$$R/C = \frac{(134)(550)}{3000} = \boxed{24.6 \text{ ft/sec}}$$

6.7 Assuming $(R/C)_{max}$ varies linearly with altitude,

$$(R/C)_{max} = a h + b$$

From the two $(R/C)_{max}$ values from problem 6.5,

$$87.3 = a(o) + b$$

$$48.5 = a(5000) + b$$

Hence,

$$b = 87.3$$

$$a = -0.00776$$

$$(R/C)_{max} = -0.00776 h + 87.3$$

To obtain the absolute ceiling, set $(R/C)_{max} = 0$, and solve for h.

$$h = \frac{87.3}{0.00776} = 11,250 \text{ m}$$

Hence,

$$\text{absolute ceiling} \approx \boxed{11.3 \text{ km}}$$

6.8 $(R/C)_{max} = a\,h + b$

From the results of problem 6.6,

$$42.5 = a(0) + b$$

$$24.6 = a(12{,}000) + b$$

Hence,

$$b = 42.5$$

$$a = -0.00149$$

$$(R/C)_{max} = -0.00149\,h + 42.5$$

To obtain the absolute ceiling, set $(R/C)_{max} = 0$, and solve for h.

$$h = \frac{42.5}{0.00149} = 28{,}523 \text{ ft}$$

absolute ceiling $\approx \boxed{28{,}500 \text{ ft}}$

COMMENT TO THE INSTRUCTOR

In the above problems dealing with a performance analysis of the twin-jet and single-engine piston airplanes, the answers will somewhat depend on the precision and number of calculations made by the student. For example, if the P_R curve is constructed from 30 points instead of the six or eight points as above, the subsequent results for rate-of-climb and absolute ceiling will be more accurate than obtained above. Some lee-way on the students' answers is therefore advised. In my own experience, I am glad when the students fall within the same ballpark.

6.9 $R = h\,(L/D)_{max} = 5000\,(7.7) = \boxed{38{,}500 \text{ ft} = 729 \text{ miles}}$

6.10 First, we have to calculate the C_L corresponding to maximum L/D.

$$\left(\frac{C_L}{C_D}\right)_{max} = \frac{\sqrt{C_{D,o}\pi\, e\, AR}}{2\, C_{D,o}} = \sqrt{\frac{\pi\, e\, AR}{4\, C_{D,o}}}$$

$$C_{D,o} = \frac{1}{4}\left(\frac{C_L}{C_D}\right)_{max}^{-2} \pi\, e\, AT = \frac{(0.25)\pi(0.7)(4.11)}{(7.7)^2} = 0.038$$

At maximum L/D, $C_{D,o} = C_{D,i} = 0.038$

Hence: $C_D = 2(0.038) = 0.076$

$$C_L = C_D\left(\frac{C_L}{C_D}\right) = (0.076)(7.7) = 0.585$$

Also:

$$\theta_{min} = \text{Arc Tan}\left(\frac{L}{D}\right)_{max}^{-1} = \text{Arc Tan}\,(0.13)$$

$$\theta_{min} = 7.4°$$

$$V_\infty = \sqrt{\frac{2\cos\theta}{\rho_\infty C_L}\left(\frac{W}{S}\right)} = \sqrt{\frac{2\cos(7.4°)}{(0.002175)(0.585)}\left(\frac{1400}{231}\right)} = \boxed{97.2 \text{ ft/sec}}$$

6.11 From Eq. (6.85),

$$\left(\frac{L}{D}\right)_{max} = \frac{(\pi\, e\, AR\, C_{D_o})^{1/2}}{2\, C_{D_o}}$$

Putting in the numbers:

$$\left(\frac{L}{D}\right)_{max} = \frac{[\pi(.9)(6.72)(0.025)]^{1/2}}{0.050} = \boxed{13.78}$$

6.12 Aviation gasoline weighs 5.64 lb per gallon,

Hence,

$$W_F = (44)(5.64) = 248 \text{ lb.}$$

Thus, the empty weight is

$$W_1 = 3400 - 248 = 3152 \text{ lb.}$$

The specific fuel consumption, in consistent units, is

$$c = 0.42/(550)(3600) = 2.12 \times 10^{-7} \text{ ft}^{-1}$$

The maximum L/D can be found from Eq. (6.85).

$$\left(\frac{L}{D}\right)_{max} = \frac{(\pi \ e \ AR \ C_{D_e})^{1/2}}{2 \ C_{D_e}} = \frac{[\pi(0.91)(6.2)(0.027)]^{1/2}}{2(0.027)} = 12.8$$

Thus, the maximum range is:

$$R = \left(\frac{\eta}{c}\right)\left(\frac{C_L}{C_D}\right)_{max} \ell n\left(\frac{W_o}{W_1}\right)$$

$$R = \left(\frac{0.83}{2.12 \times 10^{-7}}\right) (12.8) \ \ell n \left(\frac{3400}{3152}\right)$$

$$R = 3.8 \times 10^6 \text{ ft}$$

In terms of miles,

$$R = \frac{3.8 \times 10^6}{5280} = \boxed{719 \text{ miles}}$$

To calculate endurance, we must first obtain the value of $(C_L^{3/2}/C_D)_{max}$. From Eq. (6.87),

$$\left(\frac{C_L^{3/2}}{C_D}\right)_{max} = \frac{(3 \ C_{D_o} \pi \ e \ AR)^{3/4}}{4 \ C_{D_e}}$$

Putting in the numbers:

$$\left(\frac{C_L^{3/2}}{C_D}\right)_{max} = \frac{[3(0.027)\pi(0.91)(6.2)]^{1/2}}{4(0.027)} = 11.09$$

Hence, the endurance is

$$E = \left(\frac{\eta}{c}\right)\frac{C_L^{3/2}}{C_D}(2\rho_\infty S)^{1/2}(W_1^{-1/2} - W_o^{-1/2})$$

$$E = \left(\frac{0.83}{2.12 \times 10^{-7}}\right)(11.09)[2(0.002377)(181)]^{1/2}(3152^{-1/2} - 3400^{-1/2})$$

$$E = \boxed{2.67 \times 10^4 \text{ sec} = 7.4 \text{ hr}}$$

6.13 One gallon of kerosene weighs 6.67 lb. Since 1 lb = 4.448 N, then one gallon of kerosene also weighs 29.67 N. Thus,

$W_f = (1900)(29.67) = 56370$ N

$W_1 = W_o - W_f = 136960 - 56370 = 80590$ N

In consistent units,

$$c_t = 1.0 \frac{N}{(N)(hr)} = 2.777 \times 10^{-4} \text{ sec}^{-1}$$

Also, at a standard altitude of 8 km.

$\rho_\infty = 0.526$ kg/m^3

Since maximum range for a jet aircraft depends upon maximum $C_L^{1/2}/C_D$, we must use Eq. (6.86).

$$\left(\frac{C_L^{1/2}}{C_D}\right)_{max} = \frac{\left(\frac{1}{3} C_{D_e} \pi \, e \, AR\right)^{1/4}}{\frac{4}{3} C_{D_e}}$$

Putting in the numbers,

$$\left(\frac{C_L^{1/2}}{C_D}\right)_{max} = \frac{\left[\frac{1}{3}(0.032)(0.87)(6.5)\right]^{1/4}}{\frac{4}{3}(0.032)} = 15.46$$

For a jet airplane, the range is

$$R = 2\sqrt{\frac{2}{\rho_\infty S}}\left(\frac{1}{c_t}\right)\left(\frac{C_L^{1/2}}{C_D}\right)(W_o^{1/2} - W_1^{1/2})$$

$$R = 2\left[\frac{2}{(0.526)(47)}\right]^{1/2}\left(\frac{1}{2.777 \times 10^{-4}}\right)(15.46)\left[(136960)^{1/2} - (80590)^{1/2}\right]$$

$$R = 2.73 \times 10^6 \, m = \boxed{2730 \, km}$$

The endurance depends on $C_L C_D$. From Eq. (6.85),

$$\left(\frac{L}{D}\right)_{max} = \frac{\left(\pi \, e \, AR \, C_{D_e}\right)^{1/2}}{2 \, C_{D_e}} = \frac{[\pi(0.87)(6.5)(0.032)]^{1/2}}{2(0.032)} = 11.78$$

The endurance for a jet aircraft is

$$E = \frac{1}{c_t}\left(\frac{C_L}{C_D}\right)\ell n\left(\frac{W_o}{W_1}\right)$$

$$E = \frac{1}{2.777 \times 10^{-4}}(11.78) \, \ell n\left(\frac{136960}{80590}\right)$$

$$E = 22496 \, sec = \boxed{6.25 \, hr}$$

6.14
$$\frac{C_L^{3/2}}{C_D} = \frac{C_L^{3/2}}{C_{D_o} + \frac{C_L^2}{\pi e AR}}$$

$$\frac{d(C_L^{3/2}/C_D)}{dC_L} = \left[\left(C_{D_o} \frac{C_L^2}{\pi e AR}\right)\left(\frac{3}{2}C_L^{1/2}\right) - C_L^{3/2}\left(2\frac{C_L}{\pi e AR}\right)\right]\left(C_{D_o} + \frac{C_L^2}{\pi e AR}\right)^{-2} = 0$$

$$\frac{3}{2} C_L^{1/2} C_{D_o} - \frac{1}{2} \frac{C_L^{5/2}}{\pi e AR} = 0$$

$$3 C_{D_o} - \frac{C_L^2}{\pi e AR} = 0$$

Hence,

$$\boxed{C_{D_o} = \frac{1}{3} C_{D_i}}$$ This is Eq. (6.80)

To obtain Eq. (6.81)

$$\frac{C_L^{1/2}}{C_D} = \frac{C_L^{1/2}}{C_{D_o} + \frac{C_L^2}{\pi e AR}}$$

$$\frac{d(C_L^{1/2}/C_D)}{dC_L} = \left[\left(C_{D_o} \frac{C_L^2}{\pi e AR}\right)\frac{1}{2}C_L^{-1/2} - C_L^{-1/2}\left(\frac{2 C_L}{\pi e AR}\right)\right]\left(C_{D_o} + \frac{C_L^2}{\pi e AR}\right)^{-2} = 0$$

$$\frac{1}{2} C_{D_o} C_L^{-1/2} - \frac{3}{2} \frac{C_L^{5/2}}{\pi e AR} = 0$$

$$C_{D_o} - 3 \frac{C_L^{3/2}}{\pi e AR} = 0$$

$$\boxed{C_{D_o} = 3 C_{D_i}}$$

6.15 From Eq. (6.81), $C_{D_o} = 3 \, C_{D_i}$

$$C_{D_o} = 3 \, \frac{C_L^2}{\pi \, e \, AR}$$

Thus, $C_L = \left(\frac{1}{3} \pi \, e \, AR \, C_{D_o}\right)^{\frac{1}{2}}$

$$\left(\frac{C_L^{\frac{1}{2}}}{C_D}\right)_{max} = \frac{\left(\frac{1}{3} \pi \, e \, AR \, C_{D_o}\right)^{\frac{1}{4}}}{\frac{4}{3} C_{D_o}} \quad \text{This is Eq. (6.86)}$$

From Eq. (6.80), $C_{D_o} = \frac{1}{3} C_{D_i}$

$$C_{D_o} = \frac{C_L^2}{3 \pi \, e \, AR}$$

Thus, $C_L = \left(3 \pi \, e \, AR \, C_{D_o}\right)^{\frac{1}{2}}$

$$\left(\frac{C_L^{\frac{3}{2}}}{C_D}\right)_{max} = \frac{\left(3\pi \, e \, AR \, C_{D_o}\right)^{\frac{3}{4}}}{4 \, C_{D_o}} \quad \text{(This is Eq. (6.87)}$$

6.16 $AR = b^2/S$, hence $b = \sqrt{S \, AR} = \sqrt{(47)(6.5)} = 17.48 \, m$

$h = 5 \, ft = (5/3.28) \, m = 1.524 \, m$

$h/b = 1.524/17.48 = 0.08719$

$$\phi = \frac{(16 \, h/b)^2}{1 + (16 \, h/b)^2} = \frac{1.946}{2.946} = 0.66$$

$$V_{LO} = 1.2\ V_{stall} = 1.2\sqrt{\frac{2W}{\rho_\infty S\ C_{L,max}}} = 1.2\sqrt{\frac{2(103047)}{(1.255)(47)(0.8)}} = 80.3\ \text{m/sec}$$

Hence, $0.7\ V_{LO} = 56.2$ m/sec. This is the velocity at which the average force is evaluated.

$$q_\infty = \frac{1}{2}\rho_\infty V_\infty^2 = \frac{1}{2}(1.225)(56.2)^2 = 1935\ \text{N/m}^2$$

$$L = q_\infty\ S\ C_L = (1935)(47)(0.8) = 72760\ \text{N}$$

$$D = q_\infty\ S\ C_D = q_\infty\ S\left(C_{D_o} + \phi\ \frac{C_L^2}{\pi\ e\ AR}\right)$$

$$D = (1935)(47)\ [(0.032 + (0.66)(0.8)^2/\pi(0.87)(6.5)]$$

$$D = 5072\ \text{N}$$

From Eq. (6.90)

$$s_{LO} = \frac{1.44\ W^2}{g\ \rho_\infty C_{L,max}\{T - [D + \mu_r(W - L)]_{ave}\}}$$

$$= \frac{1.44\ (103047)^2}{(9.8)(1.225)(47)(0.8)\{80595 - [5072 + (0.2)(103047 - 72760]\}}$$

$$s_{LO} = \boxed{452\ \text{m}}$$

6.17 $b = \sqrt{S\ AR} = \sqrt{(181)(6.2)} = 33.5$ ft.

$h/b = 4/33.5 = 0.1194$

$$\phi = \frac{(16\ h/b)^2}{1 + (16\ h/b)^2} = \frac{3.65}{4.65} = 0.785$$

$$V_{LO} = 1.2\sqrt{\frac{2W}{\rho_\infty S\ C_{L,max}}} = 1.2\sqrt{\frac{2(3000)}{(0.002377)(181)(1.1)}} = 135\ \text{ft/sec}$$

$0.7 \, V_{LO} = 0.7 \, (135) = 94.5$ ft/sec.

$$q_\infty = \frac{1}{2} \rho_\infty V_\infty^2 = \frac{1}{2}(0.002377)(94.5)^2 = 10.6 \text{ lb/ft}^2$$

$$L = q_\infty \, S \, C_L = (10.6)(181)(1.1) = 2110 \text{ lb}$$

$$D = q_\infty \, S \, C_D = q_\infty \, S \left(C_{D_o} + \phi \, \frac{C_L^2}{\pi \, e \, AR} \right)$$

$$D = (10.6)(181)\left[(0.027 + \frac{(0.785)(1.1)^2}{\pi(0.91)(6.2)} \right] = 154.6 \text{ lb}$$

$$T = (550 \, HP_A)/V_\infty = (550)(285)/94.5 = 1659 \text{ lb}$$

$$s_{LO} = \frac{1.44 \, W^2}{g \, \rho_\infty S \, C_{L,max} \{T - [D + \mu_r (W - L)]_{ave}\}}$$

$$= \frac{1.44 \, (3000)^2}{(32.2)(0.002377)(181)(1.1)\{1659 - [154.6 + (0.2)(3000 - 2110]\}}$$

$$s_{LO} = \boxed{572 \text{ ft.}}$$

6.18 $\quad V_T = 1.3 \sqrt{\dfrac{2W}{\rho_\infty S \, C_{L,max}}} = 1.3 \sqrt{\dfrac{2(103047)}{(1.23)(47)(2.8)}} = 46.39$ m/sec

$0.7 \, V_T = 32.47$ m/sec.

$$q_\infty = \frac{1}{2} \rho_\infty V_\infty^2 = (0.5)(1.23)(32.47)^2 = 648.4 \text{ N/m}^2$$

Since the lift is zero after touchdown, $C_D = C_{D,O}$.

$$D = q_\infty \, S \, C_{D,O} = (648.4)(47)(0.032) = 975.2 \text{ N}$$

$$s_L = \frac{1.69 \, W^2}{g \, \rho_\infty C_{L,max} \left[D + \mu_r (W - L) \right]_{0.7 V_T}}$$

$$s_L = \frac{1.69\ (103047)^2}{(9.8)(1.23)(47)(2.8)[975.2 + (0.4)(103047)]} = \boxed{268\ m}$$

6.19 $V_T = 1.3\sqrt{\dfrac{2W}{\rho_\infty S\ C_{L,max}}} = 1.3\sqrt{\dfrac{2(3000)}{(0.002377)(181)(1.8)}} = 114.4$ ft/sec

$0.7\ V_T = 80.08$ ft/sec.

$q_\infty = \dfrac{1}{2}\rho_\infty V_\infty^2 = \dfrac{1}{2}\ 0.5\ (0.002377)(80.08)^2 = 7.62$ lb/ft^2

$D = q_\infty S\ C_{D,O} = (7.62)(181)(0.027) = 37.2$ lb

$$s_L = \frac{1.69\ W^2}{g\ \rho_\infty S\ C_{L,max}\left[D + \mu_r(W-L)\right]_{0.7V_T}}$$

$$s_L = \frac{1.69\ (3000)^2}{(32.2)(0.002377)(181)(1.8)[37.2 + 0.4\ (3000)]} = \boxed{493\ ft}$$

6.20 $V_\infty = 250$ mph $= 250\left(\dfrac{88}{60}\right)$ ft/sec $= 366.6$ ft/sec

$\quad = 366.6\ (0.3048)$ m/sec $= 111.7$ m/sec.

$q_\infty = \dfrac{1}{2}\rho_\infty V_\infty^2 = (0.5)(1.23)(111.7)^2 = 7673$ N/m^2

$L = q_\infty S\ C_{L,max} = (7673)(47)(1.2) = 4.328 \times 10^5$ N

$n = \dfrac{L}{W} = \dfrac{4.328 \times 10^5}{103047} = 4.2$

$$R = \frac{V_\infty^2}{g\sqrt{n^2-1}} = \frac{(111.7)^2}{9.8\sqrt{(4.2)^2-1}} = \boxed{312 \text{ m}}$$

$$w = \frac{V_\infty}{R} = \frac{111.7}{312} = \boxed{0.358 \text{ rad/sec}}$$

6.22 From Eq. (6.13)

$$T = D = q_\infty S C_D \tag{1}$$

From Eq. (6.1c)

$$C_D = C_{D,O} + \frac{C_L^2}{\pi e \, AR} \tag{2}$$

Combining (1) and (2)

$$T = q_\infty S \left[C_{D,O} + \frac{C_L^2}{\pi e \, AR} \right] \tag{3}$$

From Eq. (6.14)

$$L = W = q_\infty S C_L$$

$$C_L = \frac{W}{q_\infty S} \tag{4}$$

Substitute (4) into (3)

$$T = q_\infty S \left[C_{D,O} + \frac{W^2}{q_\infty^2 S \, \pi e \, AR} \right] = q_\infty S C_{D,O} + \frac{W^2}{q_\infty S \, \pi e \, AR}$$

Multiply by q_∞

$$q_\infty T = q_\infty^2 S C_{D,O} + \frac{W^2}{S \, \pi e \, AR}$$

or $\quad q_\infty^2 \, S \, C_{D,O} - q_\infty \, T + \dfrac{W^2}{S \, \pi \, e \, AR} = 0$

From the quadratic formula:

$$q_\infty = \dfrac{T \pm \sqrt{T^2 - \dfrac{4 \, S \, C_{D,O} \, W^2}{S \, \pi \, e \, AR}}}{2 \, S \, C_{D,O}}$$

$$q_\infty = \dfrac{\dfrac{T}{W}\left(\dfrac{W}{S}\right) \pm \dfrac{W}{S}\sqrt{\left(\dfrac{T}{W}\right)^2 - \dfrac{4 \, C_{D,O}}{\pi \, e \, AR}}}{2 \, C_{D,O}} = \dfrac{1}{2} \rho_\infty V_\infty^2$$

$$V_\infty^2 = \dfrac{\left(\dfrac{T}{W}\right)\left(\dfrac{W}{S}\right) + \dfrac{W}{S}\sqrt{\left(\dfrac{T}{W}\right)^2 - \dfrac{4 \, C_{D,O}}{\pi \, e \, AR}}}{\rho_\infty \, C_{D,O}}$$

$$V_\infty = \left[\dfrac{\left(\dfrac{T}{W}\right)\left(\dfrac{W}{S}\right) + \dfrac{W}{S}\sqrt{\left(\dfrac{T}{W}\right)^2 - \dfrac{4 \, C_{D,O}}{\pi \, e \, AR}}}{\rho_\infty \, C_{D,O}}\right]^{1/2}$$

For V_{max}: $T = (T_A)_{max}$

$$(V_\infty)_{max} = \left[\dfrac{\left(\dfrac{T_A}{W}\right)_{max}\left(\dfrac{W}{S}\right) + \dfrac{W}{S}\sqrt{\left(\dfrac{T_A}{W}\right)_{max}^2 - \dfrac{4 \, C_{D,O}}{\pi \, e \, AR}}}{\rho_\infty \, C_{D,O}}\right]^{1/2}$$

6.23 From Figure 6.2, $(L/D)_{max} = 18.5$, and $C_{D,O} = 0.015$. From Eq. (6.85)

$$\left(\frac{C_L}{C_D}\right)_{max} = \frac{(C_{D,O}\pi\, e\, AR)^{1/2}}{2\, C_{D,O}}$$

or, $\quad e = \dfrac{4\, C_{D,O}(C_L/C_D)^2_{max}}{\pi\, AR}$

$$e = \frac{4(0.015)(18.5)^2}{\pi\,(7.0)} = \boxed{0.83}$$

<u>Please note</u>: Consistent with the derivation of Eq. (6.85) where a parabolic drag polar is assumed with the zero-lift drag coefficient equal to the minimum drag coefficient, for the value of $C_{D,O}$ in this problem we read the minimum drag coefficient from Figure 6.2.

6.24 Drag: $D = q_\infty\, S\, C_D$ \hfill (1)

Power Available: $P_A = \eta\, P_s$ \hfill (2)

Also, $P_A = T_A\, V_\infty$

Hence $T_A = \dfrac{\eta\, P_s}{V_\infty}$ \hfill (3)

(a) At an altitude of 30,000 ft, $\rho_\infty = 0.00089068$ slug/ft³ and $T_\infty = 411.86°R$. The speed of sound is

$$a_\infty = \sqrt{\gamma R T_\infty} = \sqrt{(1.4)(1718)(411.86)} = 994.7 \text{ ft/sec}$$

Hence, at Mach one, the flight velocity is $V_\infty = 994.7$ ft/sec

$$q_\infty = \frac{1}{2} \rho_\infty V_\infty^2 = \frac{1}{2} (0.00089068)(994.7)^2 = 440.6 \text{ lb/ft}^2$$

The drag coefficient at Mach one, as given in the problem statement is

$$C_{D,O} \text{ (at } M_\infty = 1) = 10 [C_{D,O} \text{ (at low speed)}] = 10 (0.0211) = 0.211$$

Hence, drag at Mach one is

$$D = q_\infty \, S \, C_D = (440.6)(334)(0.211) = \underline{31,051 \text{ lb}}$$

The thrust available is obtained as follows. The engine produces 1500 horsepower supercharged to 17,500 ft. Above that altitude, we assume that the power decreases directly as the air density. At 17,500 ft, $\rho_\infty = 0.0013781$ slug/ft^3. Hence

$$HP = \frac{\rho_\infty \text{(at 30,000 ft)}}{\rho_\infty \text{(at 17,500 ft)}} (1500) = \frac{0.00089068}{0.0013781} (1500) = 969 \text{ HP}$$

From Eq. (3), above

$$T_A = \frac{\eta \, P_s}{V_\infty} = \frac{(0.3)(969)(550)}{994.7} = \underline{161 \text{ lb}}$$

Consider the airplane in a vertical dive at Mach one. The maximum downward vertical force is $W + T_A = 12,441 + 161 = 12,602$ lb. However, the drag is the retarding force acting vertically upward, and it is 31,051 lb. At Mach one, the drag far exceeds the maximum downward force. Hence, it is <u>not</u> possible for the airplane to achieve Mach one.

(b) At an altitude of 20,000 ft,

$$\rho_\infty = 0.0012673 \text{ slug/ft}^3$$

$$T_\infty = 447.43°R$$

$$a_\infty = V_\infty = \sqrt{(1.4)(1716)(447.43)} = 1036.8 \text{ ft/sec}$$

$$q_\infty = \frac{1}{2}\rho_\infty V_\infty^2 = \frac{1}{2}(0.0013672)(1036.8)^2 = 681 \text{ lb/ft}^3$$

$$D = q_\infty S C_D = (681)(334)(0.211) = \underline{47,993 \text{ lb}}$$

The thrust available is obtained as follows

$$HP = \frac{\rho_\infty(\text{at } 20,000 \text{ ft})}{\rho_\infty(\text{at } 17,500 \text{ ft})}(1500) = \frac{0.0012673}{0.0013781}(1500) = 1379 \text{ HP}$$

Hence,

$$T_A = \frac{\eta\, P_s}{V_\infty} = \frac{(0.3)(1379)(550)}{1036.8} = \underline{219 \text{ lb}}$$

In a vertical, power-on dive at 20,000 ft, the maximum vertical force downward is $W + T_A$ = 12,441 + 219 = 12,660 lb. However, the drag is the retarding force acting vertically upward, and it is 47,993 lb. At Mach one, the drag far exceeds the maximum downward force. Hence it is <u>not</u> possible for the airplane to achieve Mach 1.

7.1 $\quad C_{M_{cg}} = C_{M_{ac}} + C_L (h - h_{ac})$

$$C_{M_{ac}} = C_{M_{cg}} - C_L (h - h_{ac}) = 0.005 - 0.05 (0.03) = \boxed{-0.01}$$

7.2 $\quad q_\infty = \dfrac{1}{2} \rho_\infty V_\infty^2 = \dfrac{1}{2} (1.225)(100)^2 = 6125 \text{ N/m}^2$

At zero lift, the moment coefficient about c.g. is

$$C_{M_{cg}} = \frac{M_{cg}}{q_\infty S c} = \frac{-12.4}{(6125)(1.5)(0.45)} = -0.003$$

However, at zero lift, this is also the value of the moment coefficient about the a.c.

$$C_{M_{ac}} = \boxed{-0.003}$$

At the other angle of attack,

$$C_L = \frac{L}{q_\infty S} = \frac{3675}{(6125)(1.5)} = 0.4$$

and $\quad C_{M_{cg}} = \dfrac{M_{cg}}{q_\infty S c} = \dfrac{20.67}{(6125)(1.5)(0.45)} = 0.005$

Thus, from Eq. (7.9) in the text,

$$C_{M_{cg}} = C_{M_{ac}} + C_L (h - h_{ac})$$

Thus,

$$h - h_{ac} = \frac{C_{M_{cg}} - C_{M_{ac}}}{C_L} = \frac{0.005 - (-0.003)}{0.4}$$

$$h - h_{ac} = \boxed{0.02}$$

The aerodynamic center is two percent of the chord length <u>ahead</u> of the center of gravity.

7.3 From the results of problem 7.2,

$$q_\infty = 6125 \text{ N/m}^2$$

$$C_{M_{ac}} = -0.003$$

$$h - h_{ac} = 0.02$$

In the present problem, the c.g. has been shifted 0.2c rearward. Hence,

$$h - h_{ac} = 0.02 + 0.2 = 0.22$$

Also, $\quad C_L = \dfrac{L}{q_\infty S} = \dfrac{4000}{(6125)(1.5)} = 0.435$

Thus, $\quad C_{M_{cg}} = C_{M_{ac}} + C_L (h - h_{ac})$

$$C_{M_{cg}} = -0.003 + 0.435\,(0.22) = \boxed{0.0927}$$

7.4 From problem 7.2, we know

$$q_\infty = 6125 \text{ N/m}^2$$

$$C_{M_{ac}} = -0.003$$

$$h - h_{ac} = 0.02$$

From information provided in the present problem, at $\alpha_a = 5°$,

$$C_L = \frac{L}{q_\infty S} = \frac{4134}{(6125)(1.5)} = 0.45$$

Hence,

$$a = \frac{dC_L}{d\alpha} = \frac{0.45}{5} = 0.09 \text{ per degree}$$

Also, $V_H = \dfrac{\ell_t S_t}{c\,S} = \dfrac{(1.0)(0.4)}{(0.45)(1.5)} = 0.593$

From Eq. (7.26) in the text,

$$C_{M_{cg}} = C_{M_{ac}} + a\alpha_a\left[(h-h_{ac}) - V_H \frac{a_t}{a}\left(1-\frac{\partial \varepsilon}{\partial \alpha}\right)\right] + V_H a_t(i_t + \varepsilon_o)$$

$$C_{M_{cg}} = -0.003 + (0.09)(5)\left[(0.02) - (0.593)\left(\frac{0.12}{0.09}\right)(1-0.42)\right] + (0.593)(0.12)(2.0)$$

$$C_{M_{cg}} = -0.058$$

Hence, the moment is

$$M_{cg} = q_\infty \, S \, c \, C_{M_{cg}} = (6125)(1.5)(0.45)(-0.058)$$

$$M_{cg} = \boxed{-240 \text{ N/m}}$$

7.5 $$\frac{\partial C_{M_{cg}}}{\partial \alpha_a} = a\left[(h - h_{ac}) - V_H \frac{a_t}{a}\left(1 - \frac{\partial \varepsilon}{\partial \alpha}\right)\right]$$

where, from problems 7.4 and 7.2,

$a = 0.09$ per degree

$h - h_{ac} = 0.02$

$C_{M_{ac}} = -0.003$

$V_H = 0.593$

$a_t = 0.12$

$\dfrac{\partial \varepsilon}{\partial \alpha} = 0.42$

$i_t = 2.0°$

Thus, $\dfrac{\partial C_{M_{cg}}}{\partial \alpha_a} = (0.09)\left[(0.02) - (0.593)\left(\dfrac{0.12}{0.09}\right)(1 - 0.42)\right] = \boxed{-0.039}$

The slope of the moment coefficient curve is negative, <u>hence the airplane model is statically stable</u>. To examine whether or not the model is balanced, first calculate C_{M_o}.

$$C_{M_o} = C_{M_{ac}} + V_H a_t (i_t + \varepsilon_o) = -0.003 + (0.593)(0.12)(2.0 + 0) = 0.139$$

The trim angle of attack can be found from

$$C_{M_{cg}} = C_{M_o} + \dfrac{C_{M_{cg}}}{\partial \alpha}\alpha_e = 0$$

$$\alpha_e = -C_{M_o} / (\partial C_{M_{cg}} / \partial \alpha) = -0.139/(-0.039) = \boxed{3.56°}$$

This is a reasonable angle of attack, falling within the normal flight range. Hence, <u>the airplane model is also balanced</u>.

7.6 $\quad h_n = h_{ac_{wb}} + V_H \dfrac{a_t}{a}\left(1 - \dfrac{\partial \varepsilon}{\partial \varepsilon}\right)$

From Problem 7.2, $h_n - h_{ac_{wb}} = 0.02$

Hence,

$$h_{ac_{wb}} = h - 0.02 = 0.26 - 0.02 = 0.24$$

$$h_n = 0.24 + 0.593 \left(\frac{0.12}{0.09}\right)(1 - 0.42)$$

$$h_n = \boxed{0.70}$$

By definition,

$$\text{static margin} = h_n - h = 0.70 - 0.26 = \boxed{0.44}$$

7.7 $\quad \delta_{trim} = \dfrac{C_{M_o} + \left(\partial C_{M_{cg}} / \partial \alpha_a\right)\alpha_n}{V_H \left(\partial C_{L_t} / \partial \delta_e\right)}$

$C_{M_o} = 0.139$ (from Problem 7.5)

$\partial C_{M_{cg}} / \partial \alpha_a = -0.039$ (from Problem 7.5)

$\alpha_n = 8°$

$V_H = 0.593$ (from Problem 7.4)

$\partial C_{L_t} / \partial \delta_e = 0.04$ (given)

Thus, $\quad \delta_{trim} = \dfrac{0.139 + (-0.039)(8)}{(0.593)(0.04)} = \boxed{-7.29°}$

7.8 $\quad F = 1 - \dfrac{1}{a_t}\left(\dfrac{\partial C_{L_t}}{\partial \delta_e}\right)\dfrac{(\partial C_{he}/\partial \alpha_t)}{(\partial C_{he}/\partial \delta_e)}$

$$F = 1 - \dfrac{1}{0.12}(0.04)\dfrac{(-0.007)}{(-0.012)} = 0.806$$

$$C'_{M_o} = C_{M_{ac_{wb}}} + F\, V_H\, a_t\, (i_t + \varepsilon_o)$$

$$C'_{M_o} = -0.003 + (0.806)(0.593)(0.12)(2+0) = \boxed{0.112}$$

This is to be compared with $C_{M_o} = 0.139$ from Problem 7.5 for stick-fixed stability.

$$h'_n = h_{ac_{wb}} + F\, V_H\, \dfrac{a_t}{a}\left(1 - \dfrac{\partial \varepsilon}{\partial \alpha}\right)$$

$$h'_n = 0.24 + (0.806)(0.593)\dfrac{(0.12)}{(0.09)}(1 - 0.42) = \boxed{0.609}$$

This is to be compared with $h_n = 0.70$ from Problem 7.6 for stick-fixed stability.

$$h'_n - h = 0.609 - 0.26 = \boxed{0.349}$$

Note that the static margin for stick-free is 79% of that for stick-fixed.

$$\dfrac{\partial C'_{M_{cg}}}{\partial \alpha_a} = -a\left(h'_n - h\right) = -(0.09)(0.349) = \boxed{-0.031}$$

This is to be compared with a slope of -0.039 obtained from Problem 7.5 for the stick-fixed case.

8.1

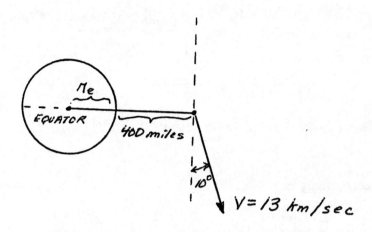

$h_G = 400$ miles $= 0.644 \times 10^6$ m

$r_b = r_e + h_G = 6.4 \times 10^6 + 0.644 \times 10^6 = 7.044 \times 10^6$ m

$k^2 = 3/986 \times 10^{14}$ m^3/sec^2 from the text.

$h = r^2 \dot{\theta} = r V_\theta = r_b V \cos \beta_b = (7.044 \times 10^6)(13 \times 10^3) \cos 10°$

$h = 9.018 \times 10^{10}$ m^2/sec

$h^2 = 8.133 \times 10^{21}$ m^4/sec^2

$p = h^2/k^2 = 8.133 \times 10^{21}/3.986 \times 10^{14} = 2.04 \times 10^7$ m

This is the numerator of the orbit equation. We now proceed to find the eccentricity.

The kinetic energy per unit mass is:

$T/m = V^2/2 = (13 \times 10^3)^2/2 = 8.45 \times 10^7$ m^2/sec^2

The potential energy per unit mass is:

$$|\phi / m| = k^2/r_b = 3.98 \times 10^{14}/7.044 \times 10^6 = 5.659 \times 10^7 \text{ m}^2/\text{sec}^2$$

Hence,

$$H/m = (T - |\phi|)/m = (8.45 - 5.659) \times 10^7 = 2.791 \times 10^7 \text{ m}^2/\text{sec}^2$$

Thus, the eccentricity is:

$$e = \sqrt{1 + \frac{2h^2}{k^4}\left(\frac{H}{m}\right)} = \sqrt{1 + \frac{2(8.133 \times 10^{21})(2.791 \times 10^7)}{(3.986 \times 10^{14})^2}} = 1.96$$

Obviously, the trajectory is a hyperbola because e > 1, and also because T > |ϕ|. The orbit equation is

$$r = \frac{p}{1 + e \cos(\theta - c)}$$

$$r = \frac{2.04 \times 10^7}{1 + 1.96 \cos(\theta - C)}$$

The phase angle, C, is calculated as follows. Substitute the burnout location ($r_b = 7.044 \times 10^6$ m and $\theta = 0°$) into the above equation.

$$7.044 \times 10^6 = \frac{2.04 \times 10^7}{1 + 1.96 \cos(-C)}$$

$$\cos(-C) = 0.967$$

Thus C = -14.76°.

Hence, the complete equation of the trajectory is

$$r = \frac{2.04 \times 10^7}{1 + 1.96 \cos(\theta + 14.76)}$$

where θ is in degrees and r in meters.

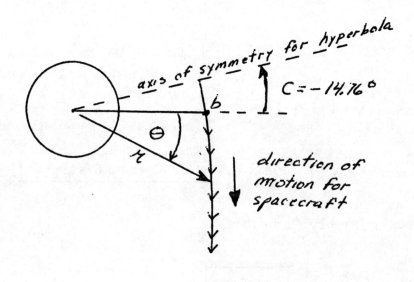

8.2 Escape velocity = $V = \sqrt{2k^2/r}$

For Venus: $V = \sqrt{(2)(3.24 \times 10^{14})/6.16 \times 10^6} = 1.03 \times 10^4$ m/sec = $\boxed{10.3 \text{ km/sec}}$

For Earth: $V = \sqrt{(2)(3.96 \times 10^{13})/6.39 \times 10^6} = 1.11 \times 10^4$ m/sec = $\boxed{11.3 \text{ km/sec}}$

For Mars: $V = \sqrt{(2)(4.27 \times 10^{13})/3.39 \times 10^6} = 5.02 \times 10^3$ m/sec = $\boxed{5.02 \text{ km/sec}}$

For Jupiter: $V = \sqrt{(2)(1.27 \times 10^{17})/7.14 \times 10^7} = 5.96 \times 10^4$ m/sec = $\boxed{59.6 \text{ km/sec}}$

8.3 $G = 6.67 \times 10^{-11}$ m^3/(kg)(sec)^2

$M = 7.35 \times 10^{22}$ kg

$GM = k^2 = (6.67 \times 10^{-11})(7.35 \times 10^{22}) = 4.9 \times 10^{12}$ m^3/sec^2

Orbital velocity is $V = \sqrt{k^2/r}$

$V_{orbital} = \sqrt{4.9 \times 10^{12}/1.74 \times 10^6} = \boxed{1678 \text{ m/sec} = 1.678 \text{ km/sec}}$

Escape velocity is larger by a factor of $(2)^{1/2}$.

$V_{escape} = \sqrt{2}\,(1.678) = \boxed{2.37 \text{ km/sec}}$

8.4 From Kepler's 3rd law,

$$\left(\frac{\tau_1}{\tau_2}\right)^2 = \left(\frac{a_1}{a_2}\right)^3$$

$a_2^3 = (a_1)^3\,(\tau_2/\tau_1)^2$

$$a_2 = a_1 (\tau_2/\tau_1)^{2/3}$$

$$a_2 = (1.495 \times 10^{11})(29.7/1.0)^{2/3} = \boxed{1.43 \times 10^{12} \text{ m}}$$

Please note: The "distant planet" is in reality Saturn.

8.5 In order to remain over the same point on the Earth's equator at all times (assuming the Earth is a perfect sphere), the satellite must have a circular orbit with a period of 24 hours = $8.64 \cdot 10^4$ sec. As part of the derivation of Kepler's third law in the test, it was shown that

$$\tau^2 = \left(\frac{4\pi^2}{k^2}\right) a^3 = \left(\frac{4\pi^2}{k^2}\right) r^3 \quad \text{(Remember, the orbit is a circle.)}$$

Hence,

$$r = \left(\frac{k^2}{4\pi^2}\right)^{1/3} \tau^{2/3}$$

$$r = \left(\frac{3.956 \times 10^{14}}{4\pi^2}\right)^{1/3} (8.64 \times 10^4)^{2/3} = 4.21 \times 10^7 \text{ m}$$

The radius of the Earth is 6.4×10^6 m. Hence, the altitude above the surface of the Earth is

$$h_G = 4.21 \times 10^7 - 6.4 \times 10^6 = 3.57 \times 10^7 \text{ m} = \boxed{35{,}700 \text{ km}}$$

Circular velocity is

$$V = \sqrt{\frac{k^2}{r}} = \sqrt{\frac{3.956 \times 10^{14}}{4.21 \times 10^7}} = \boxed{3065 \text{ m/sec}}$$

8.6 $\quad m = \rho v = \rho \left(\frac{4}{3}\pi r^3\right) = (6963)\left(\frac{4}{3}\pi\right)(0.8)^3 = 1.4933 \times 10^4 \text{ kg}$

$$S = \pi r^2 = \pi (0.8)^2 = 2.01 \text{ m}^2$$

$$\frac{m}{C_D S} = \frac{1.4933 \times 10^4}{(1)(2.01)} = 7429 \text{ kg/m}^3$$

$$Z = g_o/RT = 9.8/(287)(29\backslash 88) = 0.000118 \text{ m}^{-1}$$

From Eq. (8.97), the density at the altitude for maximum deceleration is

$$\rho = \frac{m}{C_D S} Z \sin\theta = (7429)(0.000118)\sin 30° = 0.4383 \ \frac{\text{kg}}{\text{m}^3}$$

From Eq. (8.73)

$$h = -\frac{1}{Z} \ln(\rho/\rho_o) = -\frac{1}{0.000118}\ln\left(\frac{0.4383}{1.225}\right) = \boxed{8710 \text{ m}}$$

From Eq. (8.100)

$$\left|\frac{dV}{dt}\right|_{max} = \frac{V_E^2 \, Z\sin\theta}{2e} = \frac{(8000)^2 (0.000118)(\sin 30°)}{2e} = 694.6 \text{ m/sec}^2$$

In terms of g's:

$$\left|\frac{dV}{dt}\right|_{max} = \frac{694.6}{9.8} = \boxed{70.88 \text{ g's}}$$

From Eq. (8.87)

$$V/V_E = e^{-\left[\frac{(1\,225)}{2(7429)(0\,000118)\sin 30°}\right]} = 0.247$$

$$V = 0.247\, V_E = 0.247\,(8000) = \boxed{1978 \text{ m/sec}}$$

8.7 Aerodynamic heating varies as V_∞^3. Hence:

$$\frac{q_2}{q_1} = \left(\frac{36000}{27000}\right)^3 = 2.37$$

$$q_2 = (100)(2.37) = \boxed{237 \text{ Btu/(ft}^2\text{)(sec)}}$$

9.1 Following the nomenclature in the text;

$$\frac{p_3}{p_2} = \left(\frac{V_2}{V_3}\right)^\gamma = 6.75^{1.4} = 13.0$$

$$p_3 = (13.0)(1.0) = 13.0 \text{ atm}$$

$$\frac{T_3}{T_2} = \left(\frac{V_2}{V_3}\right)^{\gamma-1} = (6.75)^{0.4} = 2.15$$

$$T_3 = (2.15)(285) = 613°K$$

The heat released per kg of fuel-air mixture is established in the text as 2.43×10^6 joule/kg. Hence,

$$T_4 = \frac{q}{c_v} + T_3 \quad \text{where } c_v = 720 \text{ joule/kg°K}$$

$$T_4 = \frac{2.43 \times 10^6}{720} + 613 = 3988°K$$

$$\frac{p_4}{p_3} = \frac{T_4}{T_3}$$

Thus, $\quad p_4 = p_3 \left(\dfrac{T_4}{T_3}\right) = (13.0)\left(\dfrac{3988}{613}\right) = 84.6 \text{ atm}$

$$\frac{p_5}{p_4} = \left(\frac{V_4}{V_5}\right)^{\gamma} = \left(\frac{1}{6.75}\right)^{1.4} = 0.069$$

$$p_5 = (84.6)(0.069) = 5.84 \text{ atm}$$

$$W_{compression} = \frac{p_2 V_2 - p_3 V_3}{1 - \gamma}$$

The volumes V_2 and V_3 can be obtained as follows

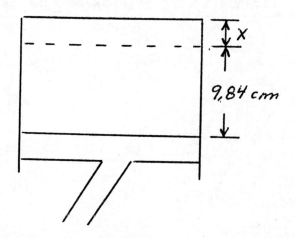

$(x + 9.84)/x = 6.75$, Thus, $x = 1.71$ cm

Thus,

$$V_2 = \frac{\pi b^2}{4}(9.84 + 1.71) \text{ where } b = \text{bore} = 11.1 \text{ cm}$$

$$V_2 = 1118 \text{ cm}^3 = 1.118 \times 10^{-3} \text{m}^3$$

$$V_3 = V_2/6.75 = 1.118 \times 10^{-3}/6.75 = 1.66 \times 10^{-4} \text{m}^3$$

Thus,

$$W_{compression} = \frac{p_2 V_2 - p_3 V_3}{1 - \gamma}$$

$$= \frac{\left[(1.0)(1.118 \times 10^{-3}) - (13.0)(1.66 \times 10^{-4})\right] 1.01 \times 10^5}{-0.4} = 263 \text{ joule}$$

$$W_{power} = \frac{p_5 V_5 - p_4 V_4}{1 - \gamma}$$

$$= \frac{\left[(5.84)(1.118 \times 10^{-3}) - (84.6)(1.66 \times 10^{-4})\right]1.01 \times 10^5}{-0.4}$$

$W_{power} = 1897$ joule

$W = W_{power} - W_{compression} = 1897 - 263 = 1634$ joule

$$P_A = \frac{1}{120} \eta \eta_{mech} (RPM) N W$$

$$P_A = \frac{(0.85)(0.83)(2800)(4)(1634)}{120} = \boxed{1.076 \times 10^5 \text{ watts}}$$

or $\quad P_A = \dfrac{1.076 \times 10^5 \text{ watts}}{746 \text{ watts/hp}} = \boxed{144 \text{ hp}}$

9.2 $\quad P_A = \dfrac{1}{120} \eta \eta_{mech} (RPM) D p_e$

where $D = \dfrac{\pi b^2}{4} s N = \dfrac{\pi (11.1)^2 (9.84)(4)}{4}$

$D = 3809$ cm^3 = 3.809×10^{-3} m^3

$$p_e = \frac{120 \, P_A}{\eta \eta_{mech} (RPM) \, D}$$

$$p_e = \frac{120(1.076 \times 10^5)}{(0.85)(0.83)(2800)(3.809 \times 10^{-3})}$$

$$p_e = \boxed{1.72 \times 10^6 \text{ N/m}^2 = 17 \text{ atm}}$$

9.3 First, calculate the mass flow through the engine. For simplicity, we will ignore the mass of the added fuel, and assume the mass flow from inlet to exit is constant. Evaluating conditions at the exit,

$$\rho_e = \frac{p_e}{RT_e} = \frac{1.01 \times 10^5}{287\,(750)} = 0.469 \text{ kg/m}^3$$

$$\dot{m} = \rho_e A_e V_e = (0.469)(0.45)(400) = 84 \text{ kg/sec}$$

At the inlet, assume standard sea level conditions.

$$\dot{m} = \rho_\infty A_i V_i = 84 \text{ kg/sec}$$

Hence,

$$V_i = \frac{\dot{m}}{\rho_\infty A_i} = \frac{84}{(1.225)(0.45)} = 152 \text{ m/sec}$$

Note: Even though the engine is stationary, it is sucking air into the inlet at such a rate that the streamtube of air entering the engine is accelerated to a velocity of 152 m/sec at the inlet. The Mach number at the inlet is approximately 0.45. Therefore, by making the assumption of standard sea level density at the inlet, we are making about a 10 percent error in the calculation of V_i. To obtain the thrust,

$$T = \dot{m}(V_e - V_i) + (p_e - p_\infty)A_e = 84(400 - 152) + (0)A_e = 20832 \text{ N}$$

Since 1 lb = 4.448 N, we also have

$$T = \frac{20832}{4.448} = \boxed{4684 \text{ lb}}$$

9.4   At a standard altitude of 40,000 ft,

$$p_\infty = 393.12 \text{ lb/ft}^2$$

$$\rho_\infty = 5.8727 \times 10^{-4} \text{ slug/ft}^3$$

The free stream velocity is

$$V_\infty = 530\left(\frac{88}{60}\right) = 777 \text{ ft/sec}$$

Hence,

$$\dot{m} = \rho_\infty V_\infty A_i = (5.87 \times 10^{-4})(777)(13) = 5.93 \text{ slug/sec}$$

$$T = \dot{m}(V_e - V_\infty) + (p_e - p_\infty)A_e$$

$$T = (5.93)(1500 - 777) + (450 - 393)(10)$$

$$T = 4287 + 570 = \boxed{4587 \text{ lb}}$$

9.5 Assume the Mach number at the end of the diffuser (hence at the entrance to the compressor) is close to zero, hence p_2 can be assumed to be total pressure.

$$M_1 = \frac{V_1}{a_1} = \frac{800}{1117} = 0.716$$

$$\frac{p_2}{p_1} = \left(1 + \frac{\gamma-1}{2} M_1^2\right)^{\gamma/\gamma-1} = [1 + (0.2)(0.616)^2]^{3.5} = 1.41$$

Hence,

$$\frac{p_3}{p_1} = \frac{p_3}{p_2} \frac{p_2}{p_1} = (12.5)(1.41) = 17.6$$

As demonstrated on the pressure-volume diagram for an ideal turbojet, the compression process is isentropic. Hence,

$$\frac{p_3}{p_1} = \left(\frac{T_3}{T_1}\right)^{\frac{\gamma}{\gamma-1}}$$

$$t_3 = t_1 \left(\frac{p_3}{p_1}\right)^{\frac{\gamma}{\gamma-1}} = 519\,(17.6)^{0.286} = 1178°$$

Combustion occurs at constant pressure. For each slug of air entering the combustor, 0.05 slug of fuel is added. Each slug of fuel releases a chemical energy (heat) of $(1.4 \times 10^7)(32.3) = 4.51 \text{ s } 10^8$ ft lb/slug. Hence, the heat released per slug of fuel-air mixture is

$$q = \frac{(4.51 \times 10^8)(0.05)}{1.05} = 2.15 \times 10^7 \text{ ft lb/slug}$$

Because the heat is added at constant pressure,

$$q = c_p (T_4 - T_3)$$

$$T_4 = T_3 + \frac{q}{c_p}$$

For air, $c_p = \dfrac{\gamma R}{\gamma - 1} = \dfrac{1.4(1716)}{0.4} = 6006 \dfrac{\text{ft lb}}{\text{slug}^\circ \text{R}}$

$$T_4 = 1178 + \frac{2.15 \times 10^7}{6006} = 1178 + 3580 = 4758^\circ \text{R}$$

Again, from the p-v diagram for an ideal turbojet,

$$\frac{p_6}{p_4} = \frac{p_1}{p_3} = \frac{1}{17.6} = 0.0568$$

Also,

$$\frac{p_6}{p_4} = \left(\frac{T_6}{T_4}\right)^{\frac{\gamma}{\gamma-1}}$$

$$T_6 = T_4 \left(\frac{p_6}{p_4}\right)^{\frac{\gamma-1}{\gamma}} = 4758 (0.0568)^{0.286} = \boxed{2095^\circ \text{R}}$$

Note: The temperatures calculated in this problem exceed those allowable for structural integrity. However, in real life, the losses due to heat conduction will decrease the temperature. Also, the fuel-air ratio would be decreased in order to lower the temperatures in an actual application.

9.6 $T = \dot{m}(V_e - V_\infty) + (p_e - p_\infty)A_e$

$1000 = \dot{m}(2000 - 950) + 0$

$\dot{m} = 0.952$ slug/sec

However,

$\dot{m} = \rho_\infty V_\infty A_i$

Thus, $A_i = \dfrac{\dot{m}}{\rho_\infty V_\infty} = \dfrac{0.952}{(0.002377)(950)} = \boxed{0.42 \text{ ft}^2}$

9.7 At 50 km, $p_\infty = 87.9$ N/m^2

$T = \dot{m}V_e + (p_e - p_\infty)A_e = (25)(4000) + (2 \times 10^4 - 87.9)(2)$

$= 100{,}000 + 39{,}824 = \boxed{139{,}824 \text{ N}}$

Since 1 lb = 4.48 N

$$T = \frac{139824}{4.448} = \boxed{31435 \text{ lb}}$$

9.8 (a) At a standard altitude of 25 km,

$p_\infty = p_e = 2527 \text{ N/m}^2$

$$I_{sp} = \frac{1}{g_o} \sqrt{\frac{2(1.18)(8314)(3756)}{(0.18)(20)} \left[1 - \left(\frac{2527}{3.03 \times 10^6}\right)^{\frac{0.18}{1.18}}\right]}$$

$I_{sp} = \boxed{375 \text{ sec}}$

(b) From the isentropic relations,

$$\frac{p_e}{p_o} = \left(\frac{T_e}{T_o}\right)^{\frac{\gamma}{\gamma-1}}$$

$$T_e = T_o \left(\frac{p_e}{p_o}\right)^{\frac{\gamma-1}{\gamma}} = 3756 \left(\frac{2527}{3.03 \times 10^6}\right)^{0.1525} = 1274°K$$

$$c_p = \frac{\gamma R}{\gamma - 1} = \frac{(1.18)(8314)}{(0.18)(20)} = 2725 \text{ joule/kg°K}$$

From the energy equation

$$V_e = \sqrt{2c_p(T_o - T_e)} = \sqrt{2(2725)(3756 - 1274)} = \boxed{3678 \text{ m/sec}}$$

(c) $\rho_e = \dfrac{p_e}{RT_e}$ where $R = \dfrac{\mathcal{R}}{M} = \dfrac{8314}{20} = 415.7 \dfrac{\text{joule}}{\text{kg}°\text{K}}$

$$\rho_e = \dfrac{2527}{(415.7)(1274)} = 0.00477 \text{ kg/m}^3$$

$$\dot{m} = \rho_e A_e V_e = (0.00477)(15)(3678) = \boxed{263.5 \text{ kg/sec}}$$

(d) From the definition of specific impulse,

$$I_{sp} = \dfrac{T}{\dot{w}} \text{ where } \dot{w} \text{ is the weight flow}$$

$$\dot{w} = \dot{m}\, g_o = (263.5)(9.8) = 2582 \text{ N/sec}$$

Hence,

$$T = \dot{w}\, I_{sp} = (2582)(375) = \boxed{968250 \text{ N}}$$

or, $T = \dfrac{968250}{4.448} = \boxed{217682 \text{ lb}}$

(e) $\dot{m} = \dfrac{p_o A^*}{\sqrt{T_o}} \sqrt{\dfrac{\gamma}{R}\left(\dfrac{2}{\gamma+1}\right)^{(\gamma+1)/(\gamma-1)}}$

$$263.5 = \dfrac{(30(1.01 \times 10^5)A^*}{(3756)^{1/2}} \sqrt{\dfrac{1.18}{415.7}\left(\dfrac{2}{2.18}\right)^{12.11}}$$

$A^* = \boxed{0.169 \text{ m}^2}$

9.9 $\dot{m} = 87.6/32.2 = 2.72$ slug/sec

$$\dot{m} = \frac{p_o A^*}{\sqrt{T_o}} \sqrt{\frac{\gamma}{R}\left(\frac{2}{\gamma+1}\right)^{(\gamma+1)/(\gamma-1)}}$$

$$2.72 = \frac{p_o^{(0.5)}}{\sqrt{6000}} \sqrt{\frac{1.21}{2400}\left(\frac{2}{2.21}\right)^{10.52}}$$

$p_o = \boxed{31736 \text{ lb/ft}^2}$

In terms of atmospheres,

$$p_o = \frac{31736}{2116} = \boxed{15 \text{ atm}}$$

9.10 $V_b = g_o I_{sp} \ln\frac{M_i}{M_f} = (9.8)(240) \ln 5.5 = \boxed{4009.6 \text{ m/sec}}$

9.11 $M_f = M_i - M_p$

$$\frac{M_i}{M_f} = \frac{M_i}{M_i - M_p} = \frac{1}{1 - \frac{M_p}{M_i}}$$

$$\frac{M_i}{M_f} = e^{V_b/g_o I_{sp}} = e^{(11,200)/(9.8)(360)} = 23.9$$

where escape velocity is 11.2 km/sec = 11,200 m/sec.

$$23.9 = \frac{1}{1 - \frac{M_p}{M_i}}$$

$$\frac{M_p}{M_i} = \boxed{0.958}$$

9.12 From Eq. (9.43)

$$r = a\, p_o^n$$

$$\log r = \log a + n \log p_o$$

At $p_o = 500$ lb/in^2: $\log (0.04) = \log a + n \log (500)$

or: $\quad \log a = -1.3979 - 2.69897\, n$ \hfill (1)

At $p_o = 1000$ lb/in^2: $\log (0.058) = \log a + n \log (1000)$

or: $\quad \log a = -1.23657 - 3n$ \hfill (2)

Solving Eqs. (1) and (1) for a and n, we have

$$0 = -0.16133 + 0.03103\,n$$

$$n = \frac{0.16133}{0.30103} = 0.5359$$

$$\log a = -1.23657 - 3(0.5359) = -2.84427$$

$$a = 0.0014313$$

Hence, the equation for the linear burning rate is

$$r = 0.0014313\, p_o^{0.5359} \tag{3}$$

For $p_o = 1500$ lb/in^2:

$$r = 0.0014313\,(1500)^{0.5359} = 0.0721 \text{ in/sec.}$$

In 5 seconds, the total distance receded by the burning surface is $0.0721\,(5) = \boxed{0.36 \text{ in.}}$

9.13

$$V_{b_1} = g_o\, I_{sp}\, \ln\left[\frac{M_{p_1} + M_{s_1} + M_{p_2} + M_{s_2} + M_L}{M_{s_1} + M_{p_2} + M_{s_2} + M_L}\right]$$

$$= (9.8)(275)\,\ln\frac{7200 + 800 + 5400 + 600 + 60}{800 + 5400 + 600 + 60}$$

$$= 1934 \text{ m/sec}$$

$$V_{b_2} - V_{b_1} = g_o\, I_{sp}\, \ln\left[\frac{M_{p2} + M_{s2} + M_L}{M_{s_2} + M_L}\right]$$

$$= (9.8)(275) \, \ell n \frac{5400 + 600 + 60}{600 + 60}$$

$$= 5975 \text{ m/sec}$$

Hence:

$$V_{b_2} = 5975 + 1934 = \boxed{7909 \text{ m/sec}}$$

10.1 $A = \dfrac{\pi d^2}{4} = \dfrac{\pi (1)^2}{4} = 0.785 \text{ in}^2$

The tensile stress is $\sigma = \dfrac{15000}{0.785} = 19{,}099 \text{ lb/in}^2$

For the AM-350 stainless steel rod:

$$\varepsilon = \frac{\sigma}{E} = \frac{19{,}099}{29 \times 10^6} = 6.586 \times 10^{-4}$$

$$\Delta \ell = \varepsilon \ell = (6.586 \times 10^{-4})(10 \text{ ft}) = 6.586 \times 10^{-3} \text{ ft.}$$

For the 2024 aluminum rod, $E = 10.7 \times 10^6$

Hence: $\Delta \ell = (6.586 \times 10^{-3} \text{ ft}) \dfrac{29}{10.7} = 17.8 \times 10^{-3} \text{ ft.}$

The aluminum rod will elongate the most, by a factor of 2.71

10.2 $A = \dfrac{\pi d^2}{4} = \dfrac{\pi (0.5)^2}{4} = 0.196 \text{ in}^2$

The yield tensile stress is 45,000 lb/in². Hence

$$F = \sigma_{ty} A = 45{,}000 \, (0.196) = \boxed{8820 \text{ lb}}$$

10.3 We first have to calculate what portion of the airplane's weight is carried by the nose wheel strut and the main gear struts. Consider the sketch below.

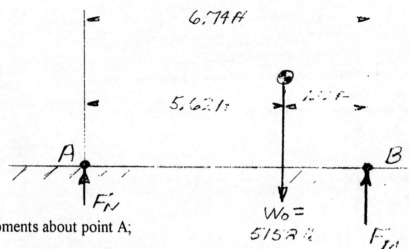

Taking moments about point A;

$$F_M = (6.74) = (5158)(5.62)$$

$$F_M = \frac{5158(5.62)}{6.74} = 4301 \text{ lb.}$$

Taking moments about point B:

$$F_N = (6.74) = (5158)(1.12)$$

$$F_N = \frac{5158(1.12)}{6.74} = 857 \text{ lb.}$$

The nose wheel strut cross section area is

$$A_N = \frac{\pi d^2}{4} = \frac{\pi (1)^2}{4} = 0.785 \text{ in}^2$$

Hence, the compressive stress in the nose wheel strut is

$$\sigma = \frac{F_N}{A_N} = \frac{857}{0.785} = \boxed{1091 \text{ lb/in}^2}$$

The main gear strut cross section area is

$$A_M = \frac{\pi d^2}{4} = \frac{\pi(3)^2}{4} = 7069 \text{ in}^2$$

The force in each main strut is $F_M/2 = 4301/2 = 2150.5$ lb. Hence the compressive stress in each main gear strut is

$$\sigma = \frac{F_M}{A_M} = \frac{2150.5}{7.069} = \boxed{304.2 \text{ lb/in}^2}$$

10.4

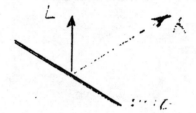

$$R = \frac{L}{\sin 60} = \frac{20}{0.866} = 23.09 \text{ lb}$$

The resultant force is balanced by the force in the cord. The cross-sectional area of the cord is

$$A = \frac{\pi d^2}{4} = \frac{\pi(0.1)^2}{4} = 0.00785 \text{ in}^2$$

Tensile stress:

$$\sigma = \frac{R}{A} = \frac{23.09}{0.00785} = \boxed{2941 \text{ lb/in}^2}$$

10.5 The shearing strain θ is given by

$$\text{Tan } \theta = \frac{0.1}{12} = 0.00833 \text{ in per in}$$

$$\theta = 0.00833 \text{ radians}$$

From Eq. (10.5): $\tau = G\theta$ $(4 \times 10^6)(0.00833) = \boxed{33{,}320 \text{ lb/in}^2}$

11.1 Since $\delta \propto \dfrac{M^2}{\sqrt{Re}}$, we have

$$\delta_{M=20} = (\delta_{M=2})\left(\frac{20}{2}\right)^2 = 0.3\,(100) = 30 \text{ inches} = \boxed{2.5 \text{ ft}}$$

This dramatically demonstrates that boundary layers at hypersonic Mach numbers can be very thick.

11.2 From Chapter 4 we have

$$\frac{T_o}{T_\infty} = 1 + \frac{\gamma - 1}{2} M_\infty^2 = 1 + \frac{1.4 - 1}{2}(20)^2 = 81.$$

At 59 km, $T_\infty = 258.1$ K

Thus: $T_o = (81)(258.1) = \boxed{20{,}906 \text{ K}}$

Considering that the surface temperature of the sun is about 6000K, the above result is an extremely high temperature. This illustrates that hypersonic flows can be very high temperature flows. At such temperatures, the air becomes highly chemically reacting, and in reality the ratio of specific heats is no longer constant; in turn, the above equation, which assumes constant γ, is no longer valid. Because the dissociation of the air requires energy (essentially "absorbs" energy), the gas temperature at the stagnation point will be much lower than calculated above; it will be approximately 6000K. This is still quite high, and is sufficient to cause massive dissociation of the air.

11.3 Following the nomenclature of Example 11.1,

$$\phi = 57.3 \, (s/R) = 57.3 \, (6.12) = 28.65°$$

$$\theta = 90° - \phi = 61.5°$$

(a)
$$\frac{p_{o_2}}{p_\infty} = \left[\frac{(\gamma+1)^2 M_\infty^2}{4\gamma M_\infty^2 - 2(\gamma-1)}\right]^{\gamma/\gamma-1} \left[\frac{1-\gamma+2\gamma M_\infty^2}{\gamma+1}\right]$$

$$= \left[\frac{(2.4)^2 (18)^2}{4(1.4)(18)^2 - 2(0.4)}\right]^{1.4/0.4} \left[\frac{1-1.4+2(1.4)(18)^2}{2.4}\right] = 417.6$$

$$C_{p,max} = \frac{2}{\gamma M_\infty^2}\left(\frac{p_{o_2}}{p_\infty} - 1\right) = \frac{2}{(1.4)(18)^2}(417.6 - 1) = \boxed{1.836}$$

(Note: This value of $C_{p,max}$ is only slightly smaller than the value calculated in Example 11.1, which was 1.838 for $M_\infty = 25$. This is an illustration of the hypersonic Mach number independence principle, which states that pressure coefficient is relatively independent of Mach number at hypersonic speeds.)

(b) From modified Newetonian:

$$C_p = C_{p,max} \sin^2\theta = (1.836) \sin^2(61.5°) = \boxed{1.418}$$

11.4 From Eq. (11.11),

$$C_L = 2 \sin^2\alpha \cos\alpha$$

$$\frac{dC_L}{d\alpha} = (2 \sin^2\alpha)(-\sin\alpha) + 4 \cos^2\alpha \sin\alpha = 0$$

$$\sin^2\alpha = 2 \cos^2\alpha = 2(1 - \sin^2\alpha)$$

$$\sin^2\alpha = 2/3$$

$$\boxed{\alpha = 54.7°}$$

$$C_{L,max} = 2 \sin^2(54.7) \cos(54.7) = \boxed{0.77}$$

11.5 (a) $C_L = 2 \sin^2\alpha \cos\alpha$

$$C_D = 2\sin^3\alpha + C_{D,o}$$

For small α, these become

$$C_L = 2\alpha^2 \tag{1}$$

$$C_D = 2\alpha^3 + C_{D,o} \tag{2}$$

$$\frac{C_L}{C_D} = \frac{2\alpha^2}{2\alpha^3 + C_{Do}} \tag{3}$$

$$\frac{d\left(\dfrac{C_L}{C_D}\right)}{d\alpha} = \frac{(2\alpha^3 + C_{D,o})4\alpha - 2\alpha^2(6\alpha^2)}{(2\alpha^3 + C_{D,o})} = 0$$

$$8\alpha^4 + 4\alpha C_{D,o} - 12\alpha^4 = 0$$

$$4\alpha^3 = 4 C_{D,o}$$

$$\alpha = (C_{D,o})^{1/3} \tag{4}$$

Substituting Eq. (4) into Eq. (3):

$$\left(\frac{C_L}{C_D}\right)_{max} = \frac{2(C_{D,o})^{2/3}}{2 C_{D,o} + C_{D,o}} = \frac{2/3}{(C_{D,o})^{1/3}} = \boxed{\frac{0.67}{(C_{D,o})^{1/3}}}$$

Hence:

$$\left(\frac{L}{D}\right)_{max} = 0.67 / (C_{D,o})^{1/3} \text{ and it occurs at}$$

$\alpha = (C_{D,O})^{1/3}$.

(b) Repeating Eq. (2):

$$C_D = 2\alpha^3 + C_{D,O} \qquad (2)$$

From the results of part (a), at $(L/D)_{max}$ we have $\alpha = (C_{,D,O})^{1/3}$. Substituting this into Eq. (2),

$$C_D = 2 C_{D,O} + C_{D,O} = 3 C_{D,O}$$

Since $C_D = C_{D,w} + C_{D,O}$

where $C_{D,w}$ is the wave drag, we have

$$C_D = C_{D,w} + C_{D,O} = 3 C_{D,O}$$

or,

$$\boxed{C_{D,w} = 2 C_{D,O}}$$

Wave drag = 2 (friction drag) when L/D is maximum. Or, another way of stating this is that friction drag is one-third the total drag.